IMAGES
of America

THE 1984 NEW ORLEANS WORLD'S FAIR

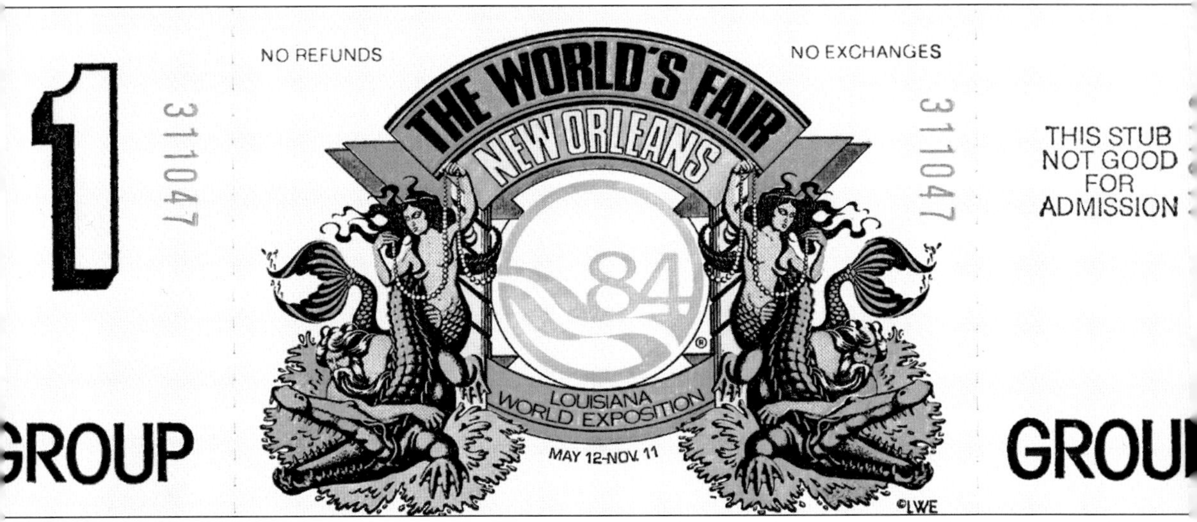

From May 12 to November 11, 1984, one of these tickets would have provided entrance to the last world's fair to be held in the United States. (Courtesy of the author.)

ON THE COVER: King Triton stands guard and a lovely mermaid relaxes with a trio of fierce alligators, all to the amusement of visitors entering the 1984 World's Fair. This was the scene at the Bridge Gate, one of the two entrances to the fair. The oversized figures helped introduce the fun to come inside while also invoking images of the famous Mardi Gras celebrations. Once safely past the imposing gate guardians, guests entered Bayou Plaza, one of the fair's six themed sections. (Courtesy of the author.)

IMAGES
of America

THE 1984 NEW ORLEANS WORLD'S FAIR

Bill Cotter

ARCADIA
PUBLISHING

ISBN 978-0-7385-6856-0

Published by Arcadia Publishing
Charleston, South Carolina

Printed in the United States of America

Library of Congress Catalog Card Number: 2008935751

For all general information contact Arcadia Publishing at:
Telephone 843-853-2070
Fax 843-853-0044
E-mail sales@arcadiapublishing.com
For customer service and orders:
Toll-Free 1-888-313-2665

Visit us on the Internet at www.arcadiapublishing.com

To Carol. Thank you for your understanding of all of the hours spent on the computer and phone while working on my research and writing, for your invaluable suggestions on the manuscript, and most of all, for your love and support all these years.

Contents

ACKNOWLEDGMENTS

This book would not have been possible without the kind assistance of a great number of individuals. A special thanks to those who worked at the fair and shared their memories of its creation and operation, especially Petr L. Spurney (president and chief executive officer of the fair), Mark Romig (the fair's director of protocol and guest relations), Daniel "Bud" Oakey (manager of Corporate Pavilions), Barry Howard (designer of the Petroleum Industries and United States pavilions), and Peggy Scott Laborde (on-site television producer/host for WDSU). They were invaluable in helping put the story of the fair into perspective, and their time is greatly appreciated.

Most of the photographs in the book are from my own collection, but I had many views of some areas and none of others. I am in debt to those who allowed the use of their own photographs to fill in those holes: James Pilet, Hap Owen, Barry Howard, David Nicholson, Eric Bouler, Ivan S. Abrams, Mike Hoffman, Ron Magill, Lee Taylor, Edward R. Cox, and Mitchel Osborne. Thanks also to the photo staffs at the Ernest N. Morial Convention Center (Rosalie Mortillaro), NASA (Margaret Persinger, Berta Alfonso, and Jeanne Ryba), the Port Of New Orleans (Donn Young and Matthew Gresham), New Orleans Public Library (Irene Wainright), and Jackson Hill Photography.

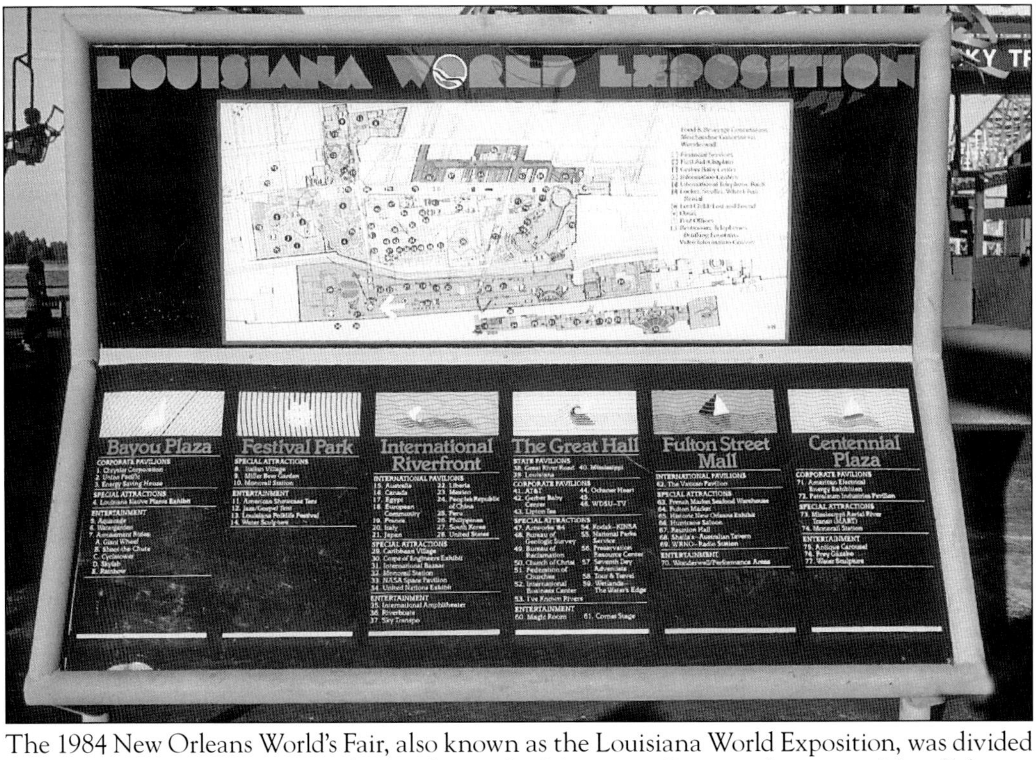

The 1984 New Orleans World's Fair, also known as the Louisiana World Exposition, was divided into six sections on an 84-acre plot bordering the Mississippi River in downtown New Orleans.

INTRODUCTION

As a fan of world's fairs, the 1984 New Orleans World's Fair has long fascinated me. It was widely panned while it was open, and many people blame its financial problems for ending the possibility of future such events in the United States. Indeed, it was the last world's fair to be held in America. And yet, despite the problems the fair encountered, it has a host of loyal fans who still recall it fondly today.

To say the fair lost money is undoubtedly true, for it ended up awash in a sea of red ink. It has the dubious distinction of being the only fair forced to declare bankruptcy while the gates were still open. Many of its participants lost large sums of money, and a long list of creditors went unpaid. Fair executives were forced to defend themselves in a long series of lawsuits that went on for years. Despite these problems, the fair has to be properly measured against the test of time.

To do so, one must first look back at the years that led up to the fair and understand the economic and political forces that brought it to life. An event as large as a world's fair just doesn't come into being without a reason, and it should not come as a surprise that the reason for this fair was money. Specifically, from the start, the fair was intended as a means of bringing more tourists—and their money—to town.

When the first talk of a world's fair surfaced in 1974, the potential exposition was touted as a way to garner sufficient state and federal funds to build a new convention center. While New Orleans has long been famous for the annual Mardi Gras celebrations, when thousands of tourists arrive to enjoy the parades, costumes, and partying, the city had failed to develop much in the way of a steady year-round tourist or convention business. New Orleans had a lot to offer, including some well-known restaurants and nightclubs, but the infrastructure simply was not there to support a significant convention trade or to attract families interested in seeing the city or the rest of Louisiana.

This was the reason that New Orleans would eventually come to host a world's fair. The fair designers built their entire project around the new convention center, for there could not be one without the other. In doing so, they brought new life into the run-down area surrounding the site. They also threw a party the likes of which most cities could only dream about.

For the six months of the fair, New Orleans had a special reason to celebrate. The fair offered visitors a variety of pavilions and shows, which is expected at any world's fair, but more importantly, it provided a condensed version of some of the excitement that is New Orleans. Visitors to the fair could enjoy live music in several different venues, ranging from small stages to a massive new amphitheater. There were daily parades capped by a fireworks show each evening, but perhaps most importantly, the fair offered some unique dining and drinking experiences to local visitors and those new to the area.

It's this feeling of fun and entertainment that made me want to tell the complete story of the fair. It wouldn't be right to gloss over the problems of the fair, but it would be worse to focus on them and ignore the fact that millions of guests came and enjoyed themselves. That is the reason that most people go to fairs—not to be educated, but to try a new experience or two and have fun in the process.

Petr Spurney, the man who ran the fair, mentioned that they had seriously miscalculated how often season pass holders would come to the fair. The official estimate of 5 to 10 uses turned out to be quite wrong; it was not uncommon for pass owners to come back 30 to 40 times. While the rest of the world may not have come in the numbers hoped for, many local residents did make the most of the event. They kept coming back because they were having fun.

Now, almost 25 years after the fair closed, and after the last of the lawsuits was settled, numerous critics have started reevaluating the fair and its impact on the city. The fair is now credited with increasing the convention and tourism businesses as originally planned, but it is also now widely acknowledged as having been the impetus behind the revitalized surrounding community. A lively arts district has taken hold, with pricey lofts replacing the little-used warehouses that were there before the fair. The city not only has its convention center but a busy cruise ship terminal, a successful mall, and numerous hotels, all on the former fair site.

This is the story I have tried to convey. The 1984 World's Fair had a lasting effect on the city, one far greater than most realized at the time. It was also, despite the notoriously hot and muggy New Orleans weather, a very special place that deserves another look. Hopefully this collection of photographs will help recapture some of that spark and magic.

– Bill Cotter
August 2008
www.worldsfairphotos.com

One

Bringing the World to New Orleans

For six months in 1984, the 82-acre site of the Louisiana World Exposition will become a separate and fanciful world.

—Pre-fair marketing booklet

When the gates of the 1984 World's Fair opened on May 12, it was the culmination of an effort that had taken 10 years and millions of dollars. The project had started in 1974, when Edward Stagg, the executive director of the Council for a Better Louisiana, saw a fair as an opportunity to build a more robust convention business. Other cities, including Seattle, Montreal, and San Antonio, had used their fairs to build the infrastructure needed to attract more visitors or to garner worldwide publicity about their existing facilities. With the success of those fairs, how could the plan fail for Louisiana?

The effort got off to a formal start in 1976 with the incorporation of the grandly named Louisiana World Exposition, Inc. The city chosen to host the fair was New Orleans, which was already the state's major tourist destination, due mainly to the famous French Quarter and the annual Mardi Gras celebrations. Hosting a fair there would be the reason for building a new convention center, which would then be used to attract conventions and their money that would inevitably flow to local hotels, restaurants, and other businesses across the state.

Organizing, building, and running a venture as ambitious as a world's fair would require a significant effort. Looking to gain credibility and to hopefully avoid mistakes made by past fairs, the organizers hired Petr Spurney, who had previously led Spokane's Expo 74, the Bicentennial's Freedom Train, and the 1980 Winter Olympics. Spurney soon found himself immersed not only in the challenging logistics required to create a world's fair, but also having to deal with a myriad of local, state, and national politics.

The organizers had originally hoped to hold the fair in 1982, but they yielded that date to the city of Knoxville for its own world's fair, a decision that proved to be a costly mistake when that event siphoned off potential exhibitors and guests. They also had to deal with a number of challenges posed by the site chosen for the fair, a rundown plot of land near the Port of New Orleans. How Spurney and his team overcame all these obstacles to open their fair is quite a tale indeed.

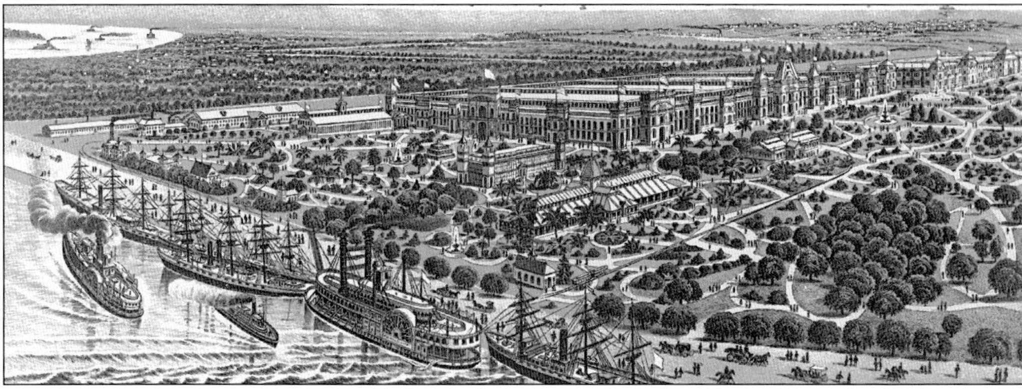

The World's Industrial and Cotton Centennial of 1884 was New Orleans's first major exposition, with a main exhibit hall that was then the largest roofed structure ever built. The exposition was plagued with construction, weather, and financial problems, and it would be another 100 years before the city would attempt another such event.

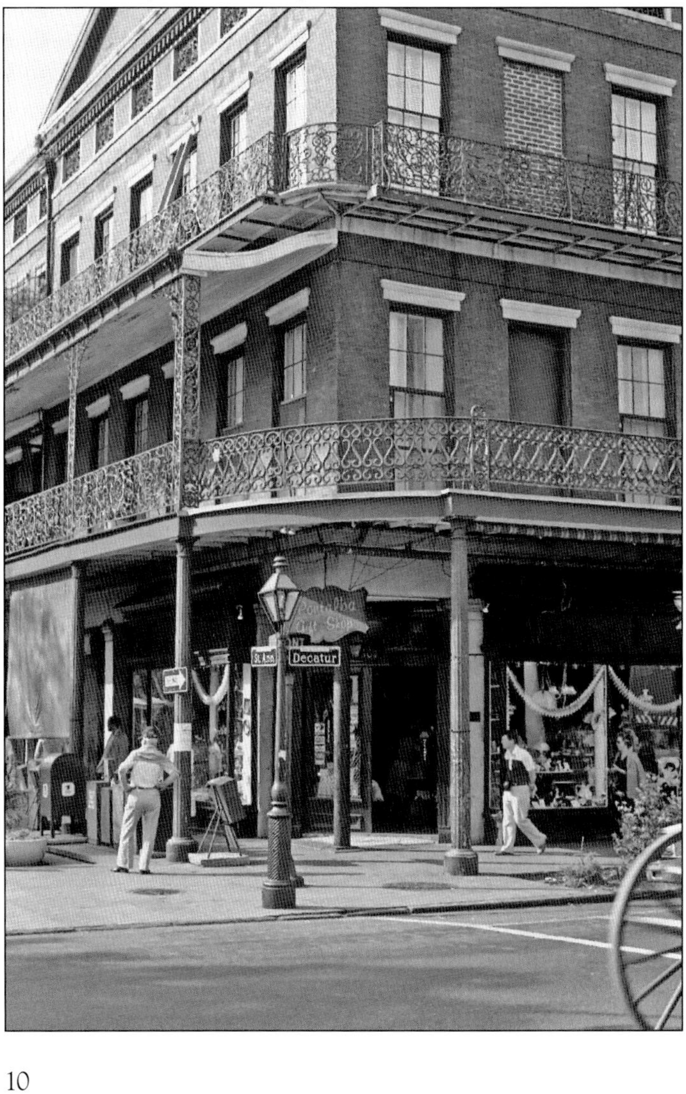

While the French Quarter section of the city was well known for the Mardi Gras parades and festivities, New Orleans civic and business leaders saw the fair as an opportunity to help turn the city into more of a year-round destination for well-healed business travelers.

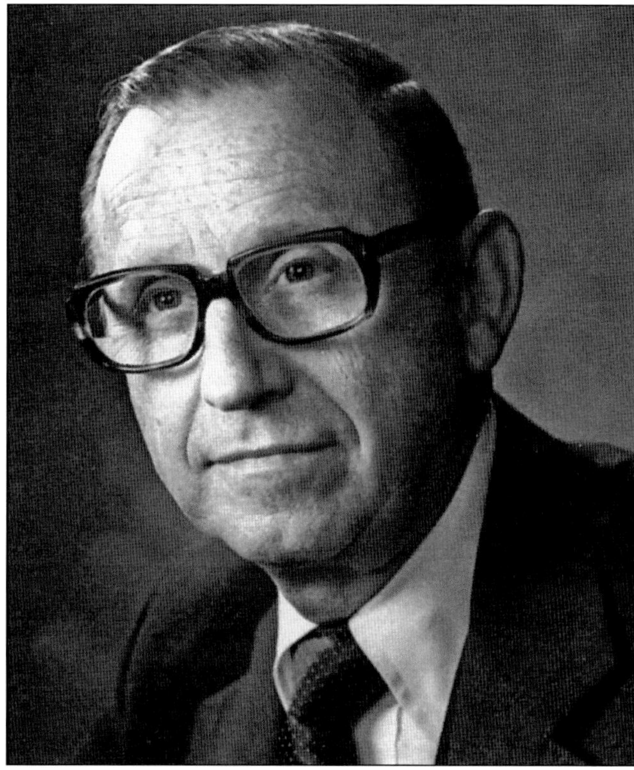

Floyd W. Lewis, a longtime local business leader and the president and chief executive officer of Middle South Utilities, was the chairman of Louisiana World Exposition, Inc. He played a key role in negotiating local and state politics to win backing for the fair.

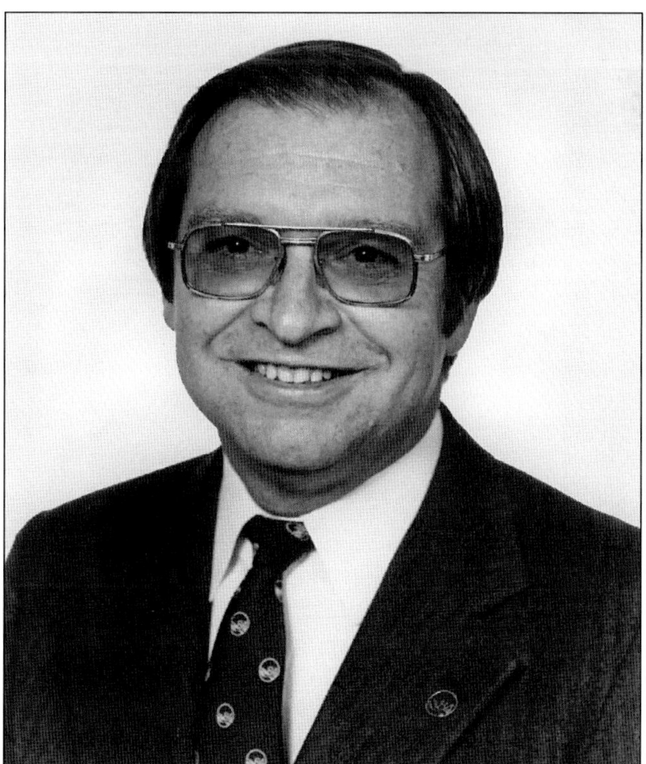

Petr L. Spurney, the fair's president and chief executive officer, was hired to build and operate the fair based on his experience leading other large-scale events. Spurney found himself enmeshed in a long series of political battles to get the fair opened and then had to successfully defend himself after the fair lost money.

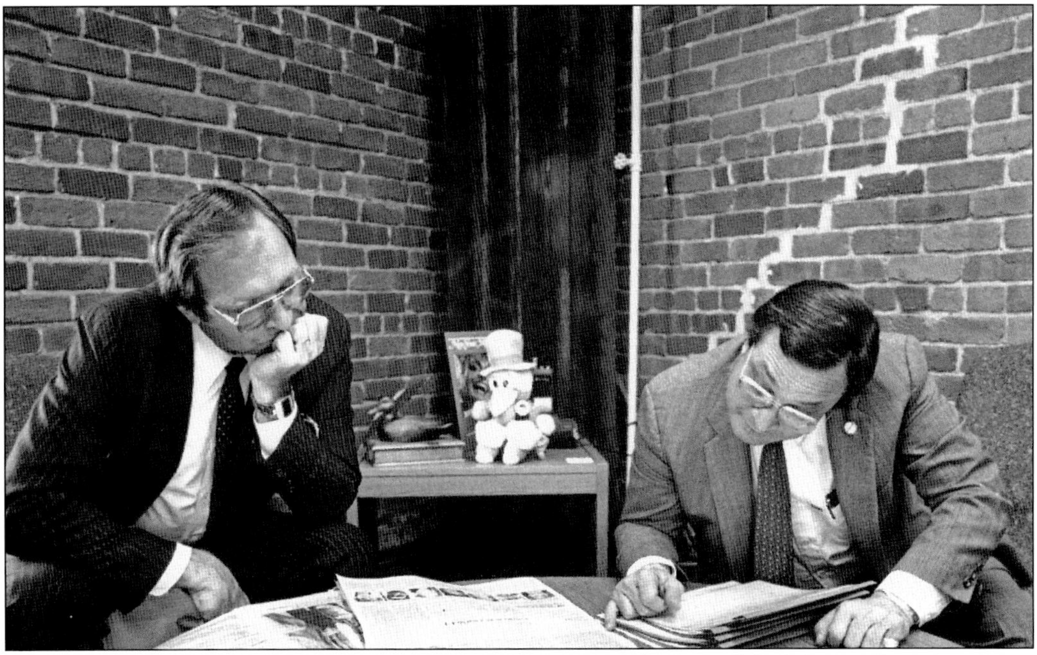

The fair also required the cooperation of Mayor Ernest "Dutch" Morial (right), seen here reviewing press coverage with Petr Spurney (left). Morial became a vocal critic of the fair when the early attendance numbers showed it was unlikely to make a profit and he launched a series of legal attacks on the fair and its officials that almost led to an early closure.

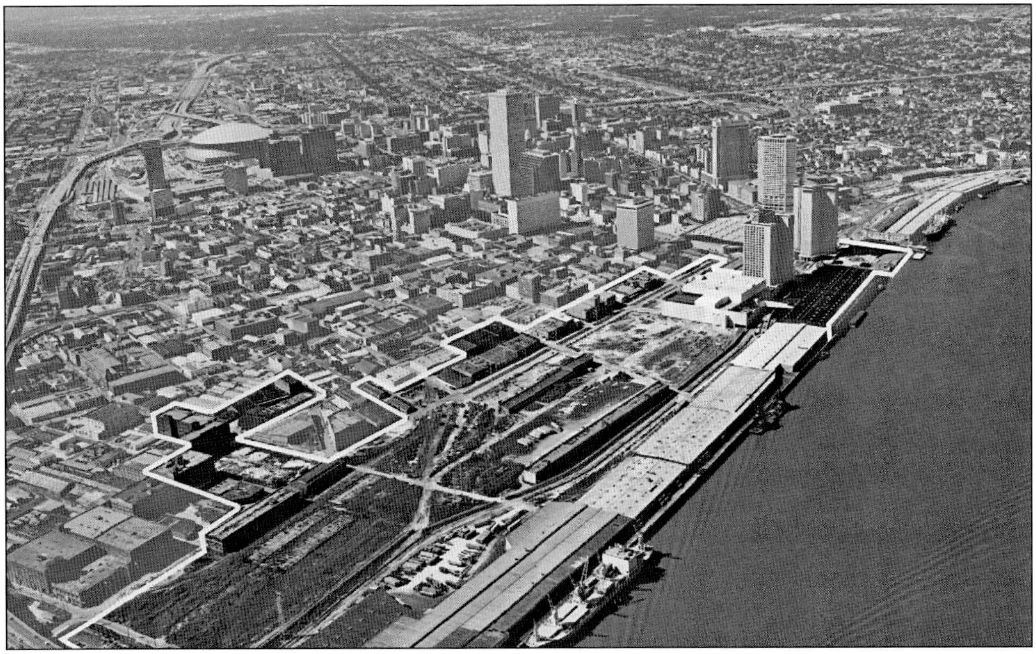

Seen just as construction began, the fair site was patched together from several parcels of real estate that included run-down warehouses, little-used wharfs, and empty plots of land. The area was ripe for revitalization, and the fair was planned as a catalyst for economic growth for the entire city.

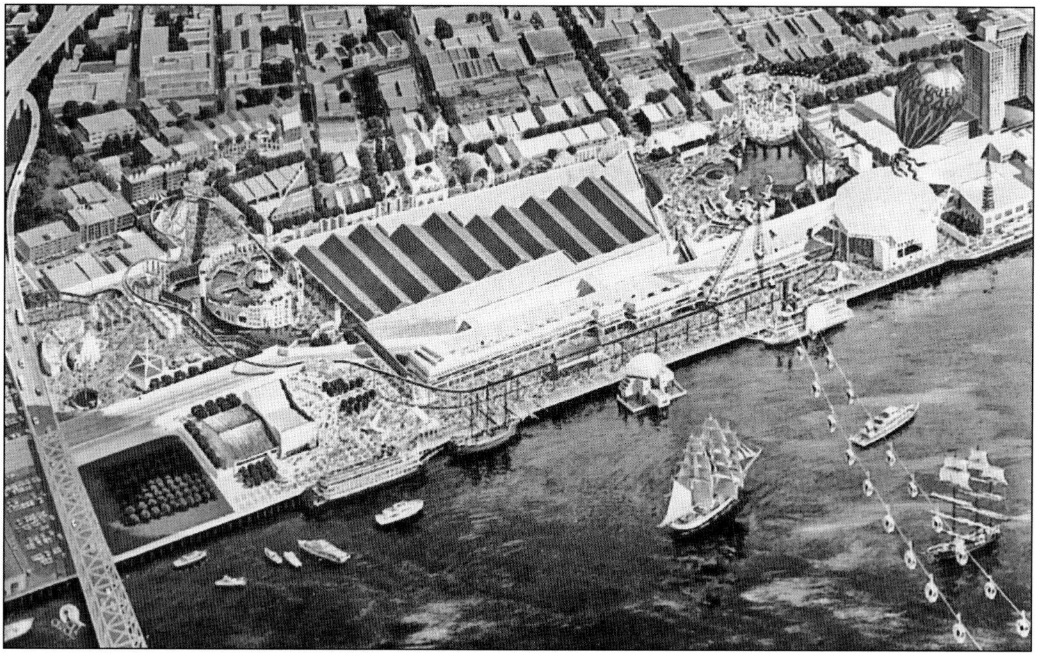

The main elements of the fair master plan were a large permanent structure that would later be used as a convention center, seen in the center of this concept art, and a riverfront structure to be converted into a shopping mall. The rest of the site would consist of temporary buildings to be removed at the end of the fair or converted warehouses near the edge of the property.

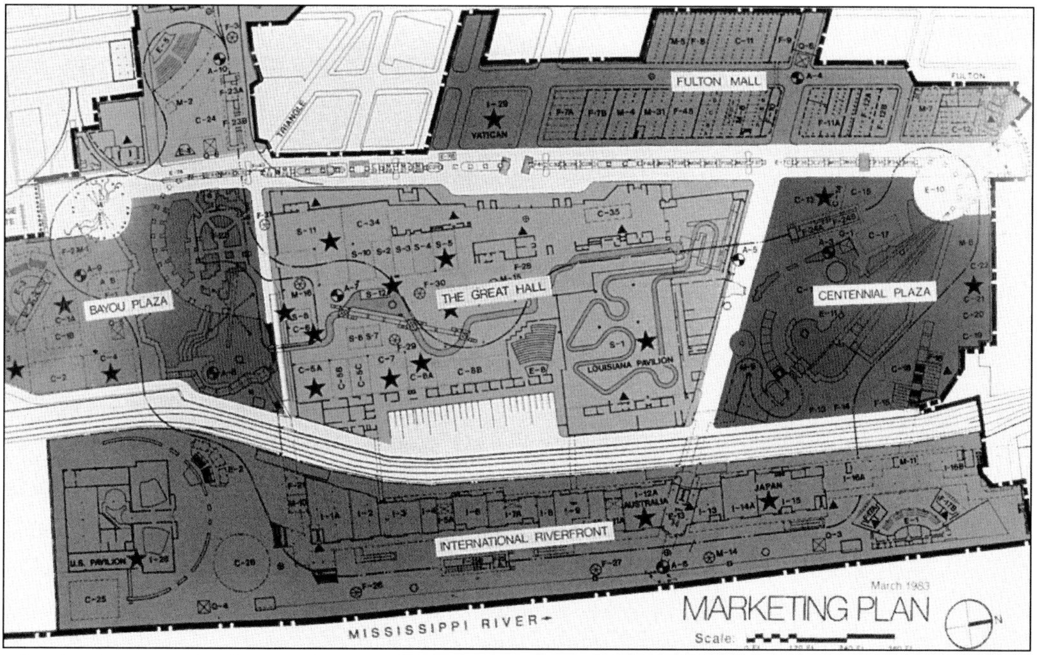

The fair designers had to work around existing streets, rail lines, power lines, and other obstacles, resulting in a site plan that divided the fair into six different sections. This enabled them to break up the look of the fair into several themed areas and helped make the site look larger than it was. Bridges, walkways, and a monorail were used to connect the areas.

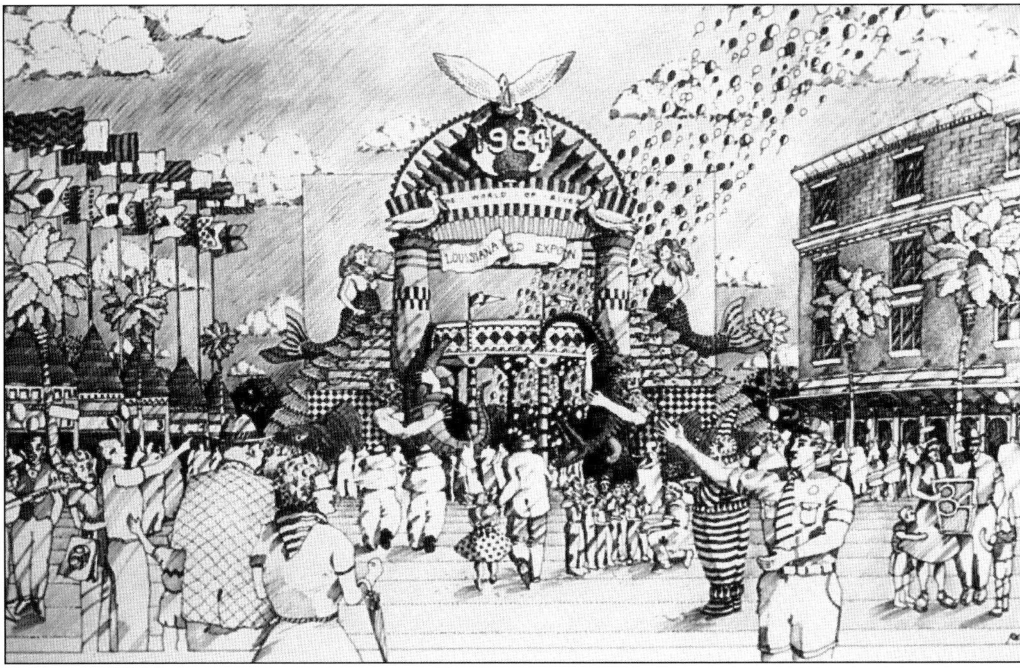

The planners produced an extensive amount of marketing material to help attract potential exhibitors, investors, and guests. This was one of the original designs for the City Gate, one of the two entrances into the fairgrounds. Designed with elements that would have been at home at Mardi Gras, the City Gate would become one of the iconic images of the fair.

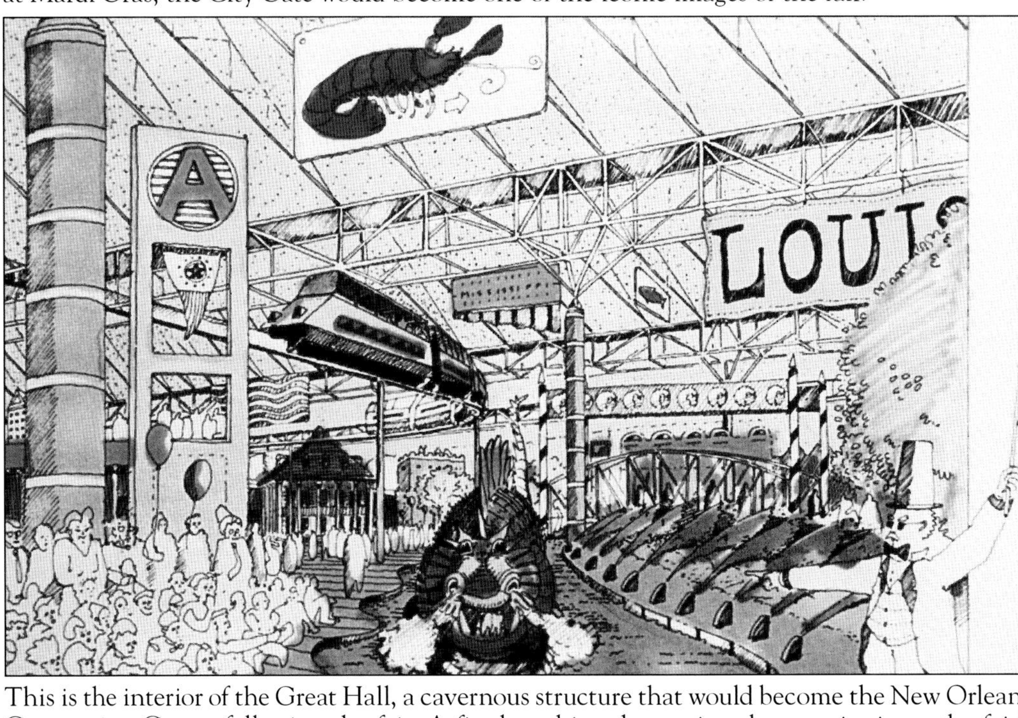

This is the interior of the Great Hall, a cavernous structure that would become the New Orleans Convention Center following the fair. A fire-breathing dragon in a lagoon ties in to the fair's theme of "The World of Rivers" as a monorail passes overhead.

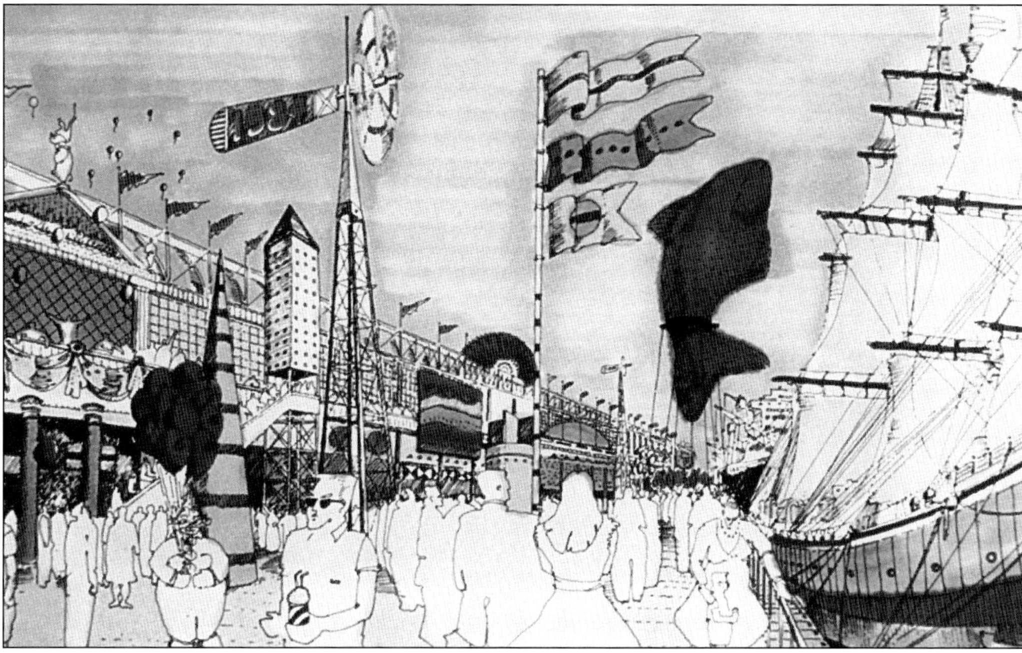

All of the foreign pavilions would be located in the International Riverfront section along the Mississippi River, where ships from various nations would be available for tours. The area would include restaurants and shops featuring specialties from the exhibiting countries as well as tourist and business information centers.

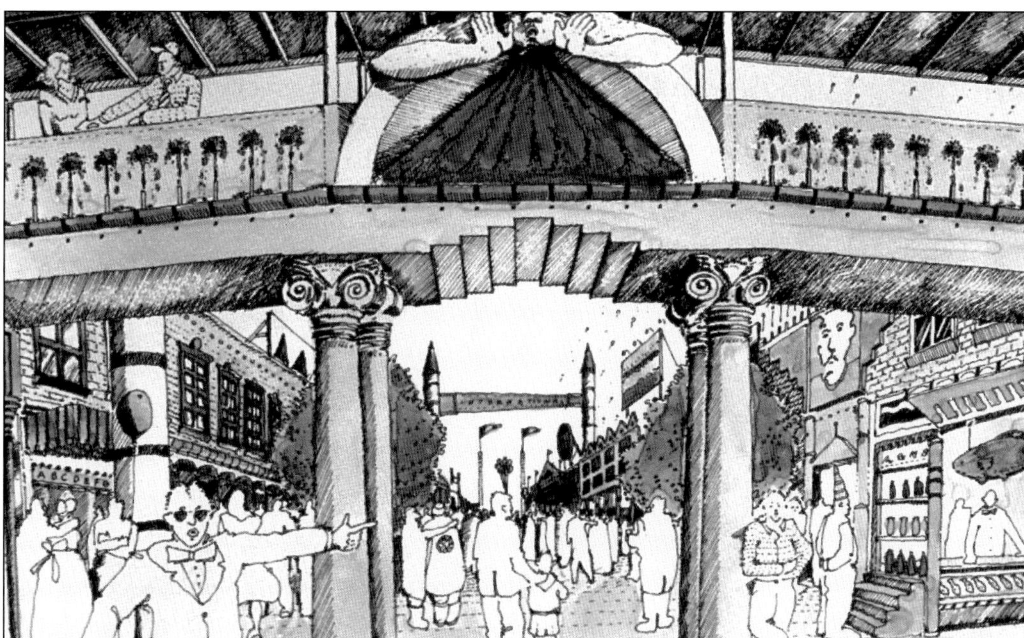

The Fulton Street Mall was one of the more challenging designs of the fair, for this area consisted of run-down warehouses that could not be razed due to their historical significance. Instead the designers would have to find a way to dress up the area and provide some color and other fun touches to the decidedly industrial structures.

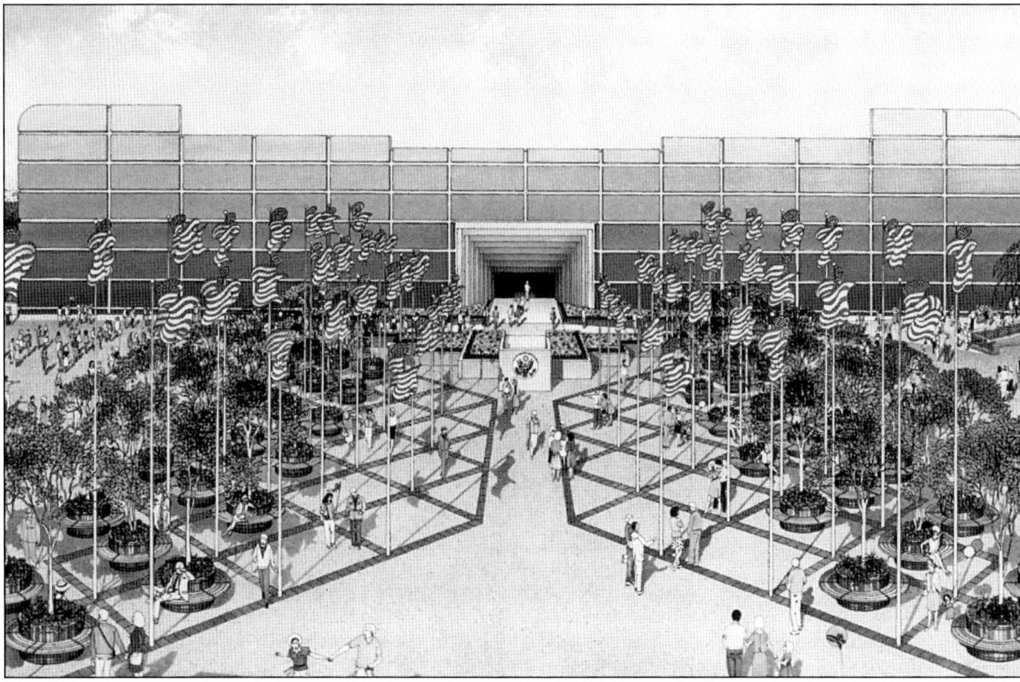

Each of the fair participants had to go through a lengthy design process for its own pavilion. The design generally began with artist renditions, such as this detailed painting of the plaza and entrance of the United States Pavilion. After the fair corporation approved these conceptual designs, work could begin on the architectural plans.

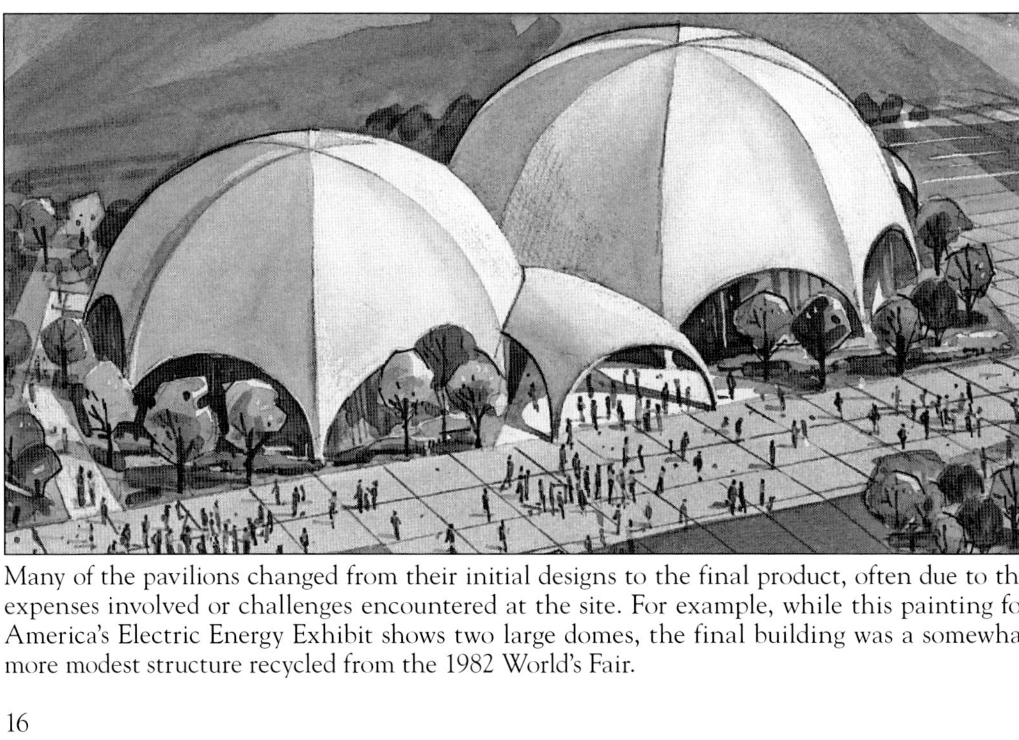

Many of the pavilions changed from their initial designs to the final product, often due to the expenses involved or challenges encountered at the site. For example, while this painting for America's Electric Energy Exhibit shows two large domes, the final building was a somewhat more modest structure recycled from the 1982 World's Fair.

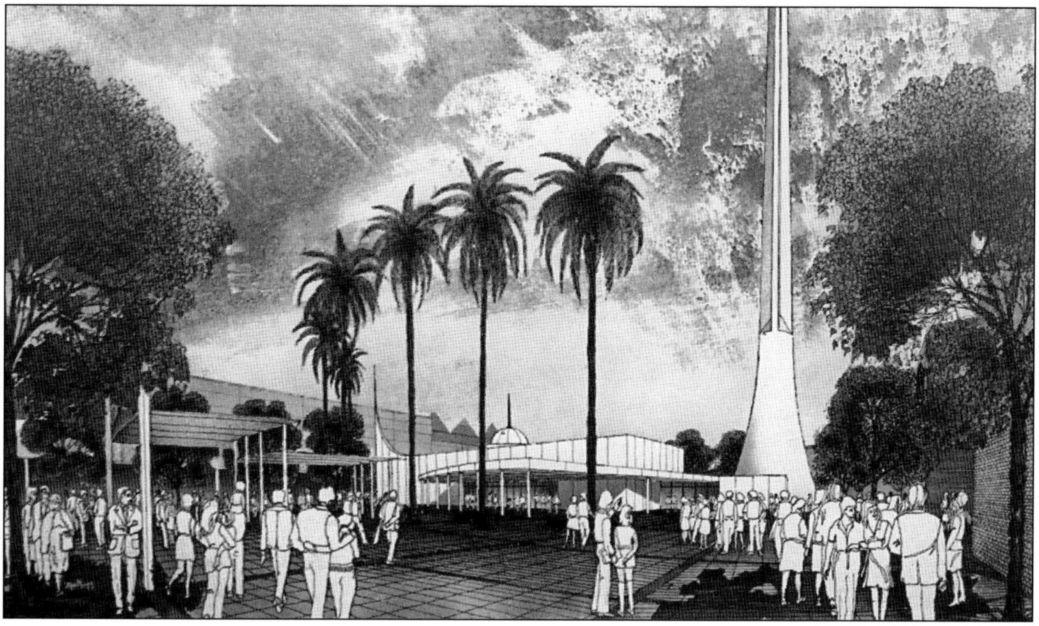

The original design for the Vatican Pavilion included a soaring tower that would have been one of the tallest structures at the fair. While the design looked attractive in this drawing, the tower would have been intrusive in its intended setting. The final version featured a much more modest cross set on a lattice-work frame.

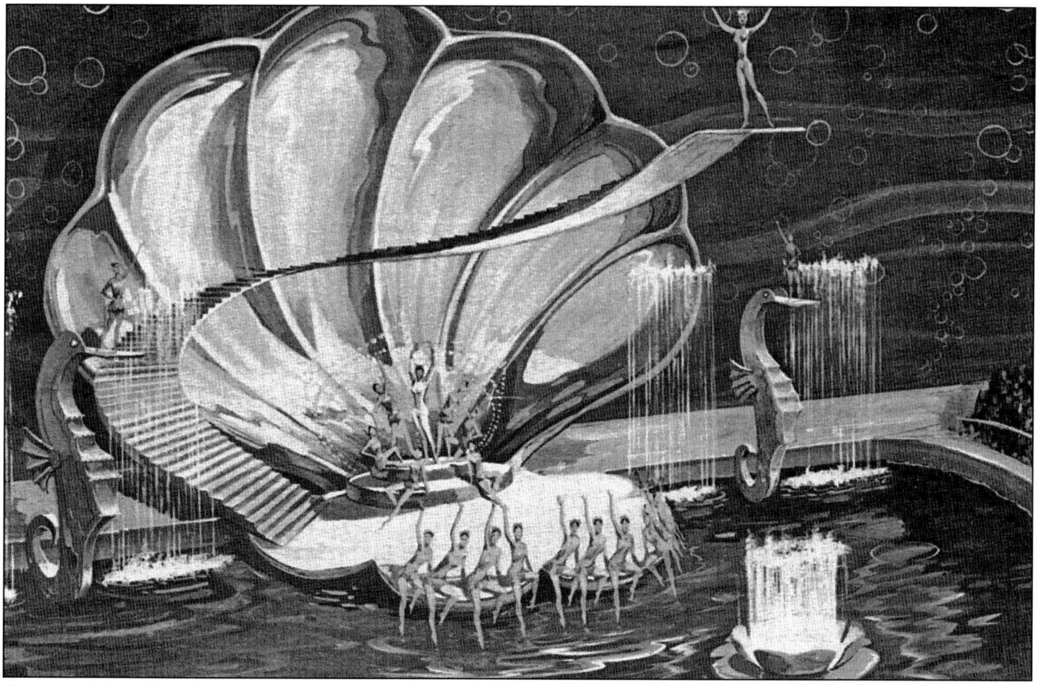

Some of the abandoned designs would have been quite a sight to behold, such as this early design for the Aquacade. The giant sea horses were to be used as diving boards with an even higher diving platform appearing between them at the end of an impossibly long staircase. The final version was a much less ambitious, but also less expensive structure.

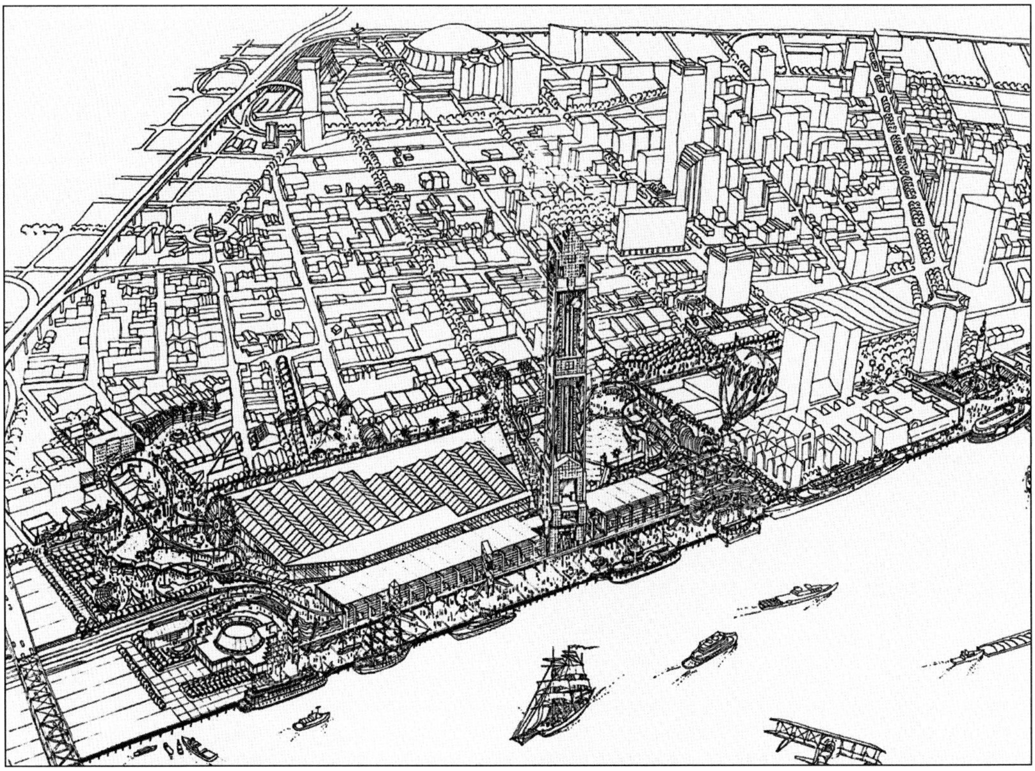

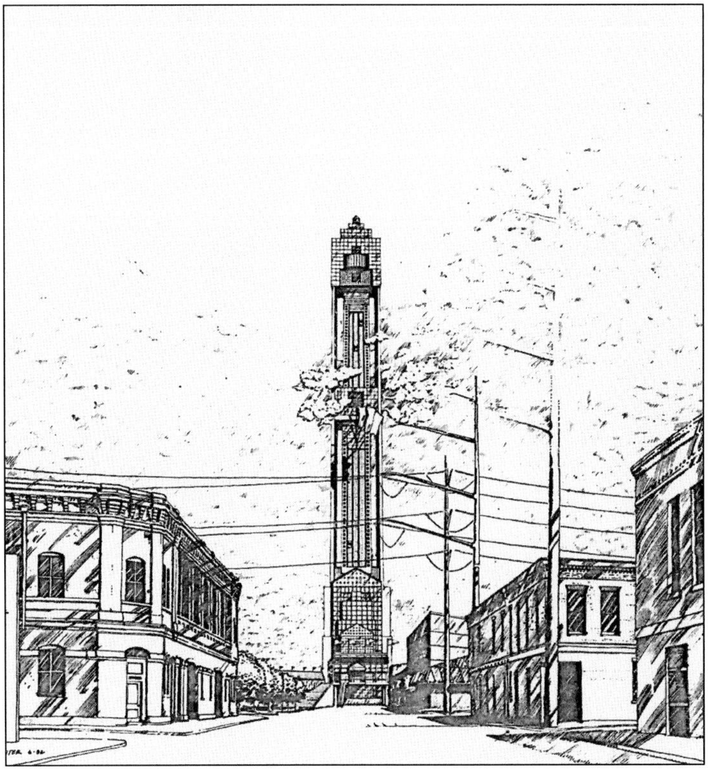

Most world's fairs have a theme structure, such as the Eiffel Tower, the Space Needle, or the Unisphere. For this fair, it was to be the 800-foot-high Tower of New Orleans. Planned as a permanent structure, it would include a rotating restaurant, interior and exterior observation decks, and an IMAX theater.

Here is how the tower would have looked from Julia Street outside the fair. While negotiations got far enough that Federal Express had agreed to lease the IMAX theater for the fair, the project could not secure adequate funding and had to be dropped from the fair design.

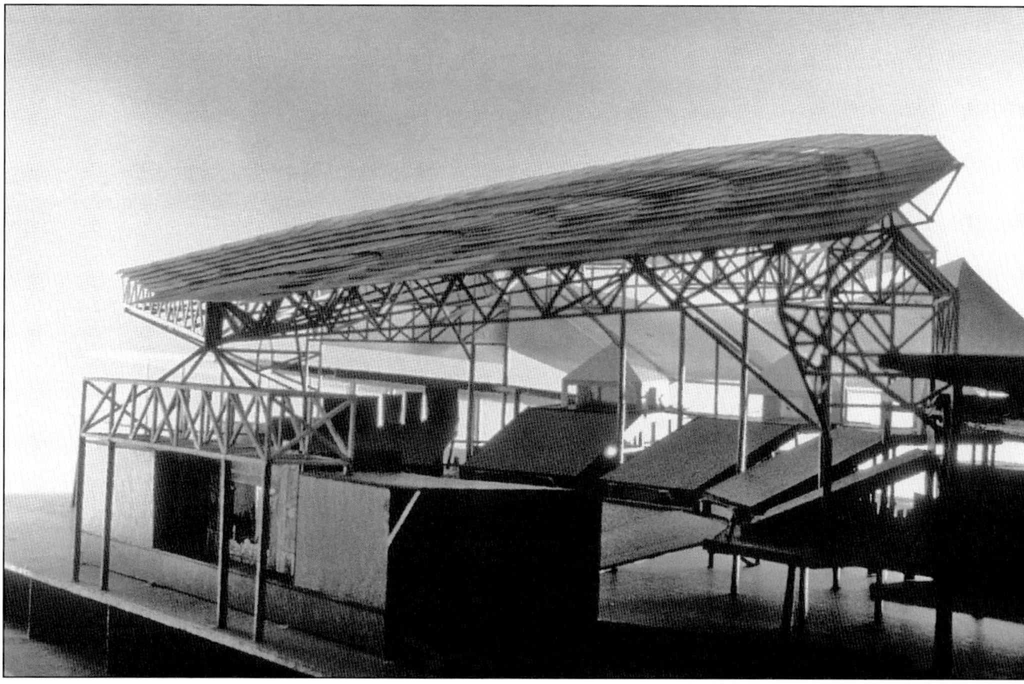

After approval of the artist rendering, the next step was usually to build a scale model of the pavilion, such as this early version of the Liggett and Myers Quality Seal Amphitheatre. These models were useful in determining sight lines and for resolving scale issues. As with the artwork, they were valuable tools in selling each project.

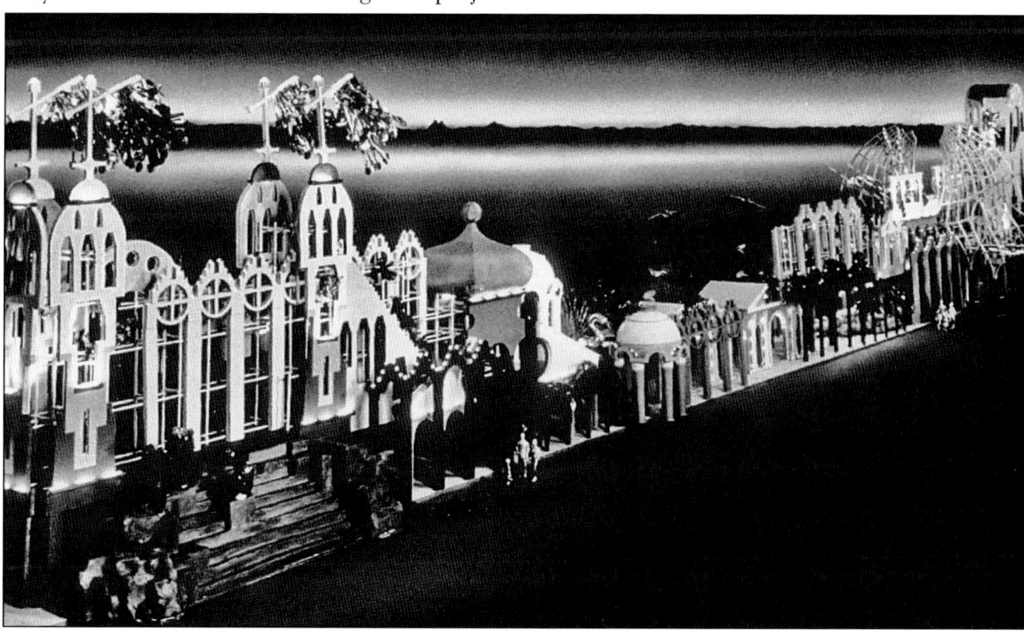

Some models, such as this one of the Wonderwall, could also be lit to study how the structure would appear at night. Others were constructed in moveable segments so designers could experiment with different configurations and layouts. A large model of the complete fair used smaller models of each of the pavilions and was used for design, marketing, and sales purposes.

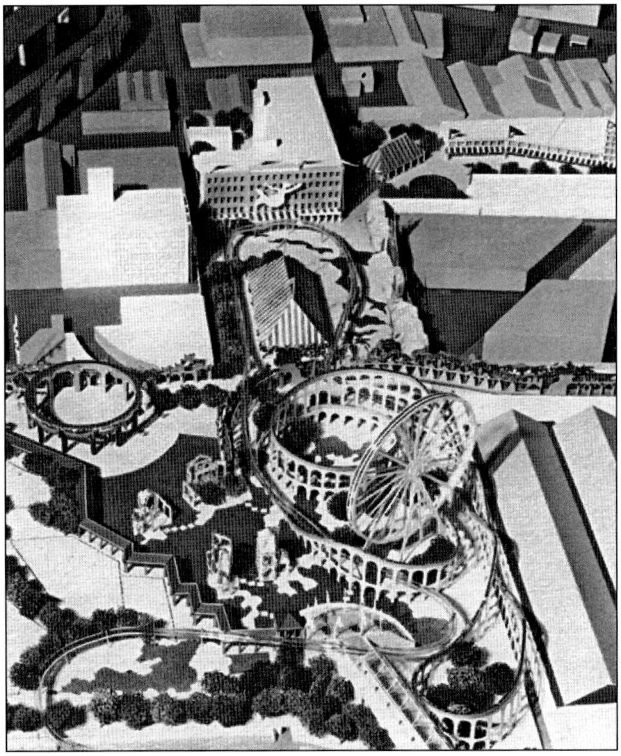

Just because a project made it into the model stage didn't mean it was guaranteed to be at the fair. This model of the Bayou Plaza section shows an elaborate roller coaster that was never actually built. The area was instead used for a series of fountains called the Watergarden.

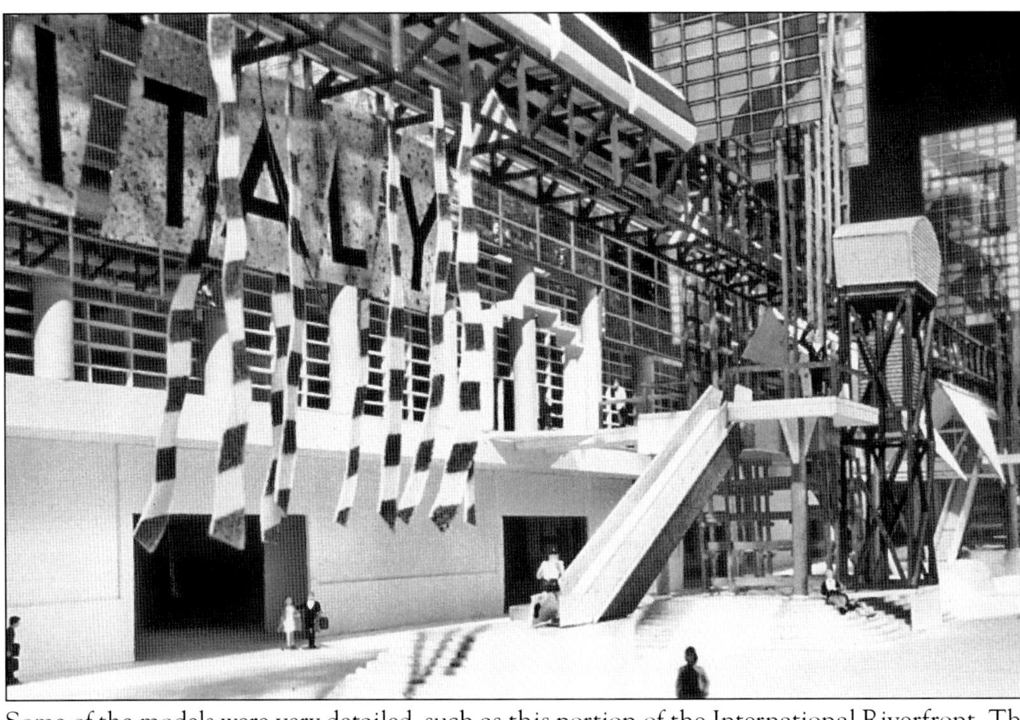

Some of the models were very detailed, such as this portion of the International Riverfront. The addition of tiny human figures helped to convey the scale of the project, as well as to make the project seem friendlier and less sterile in publicity materials.

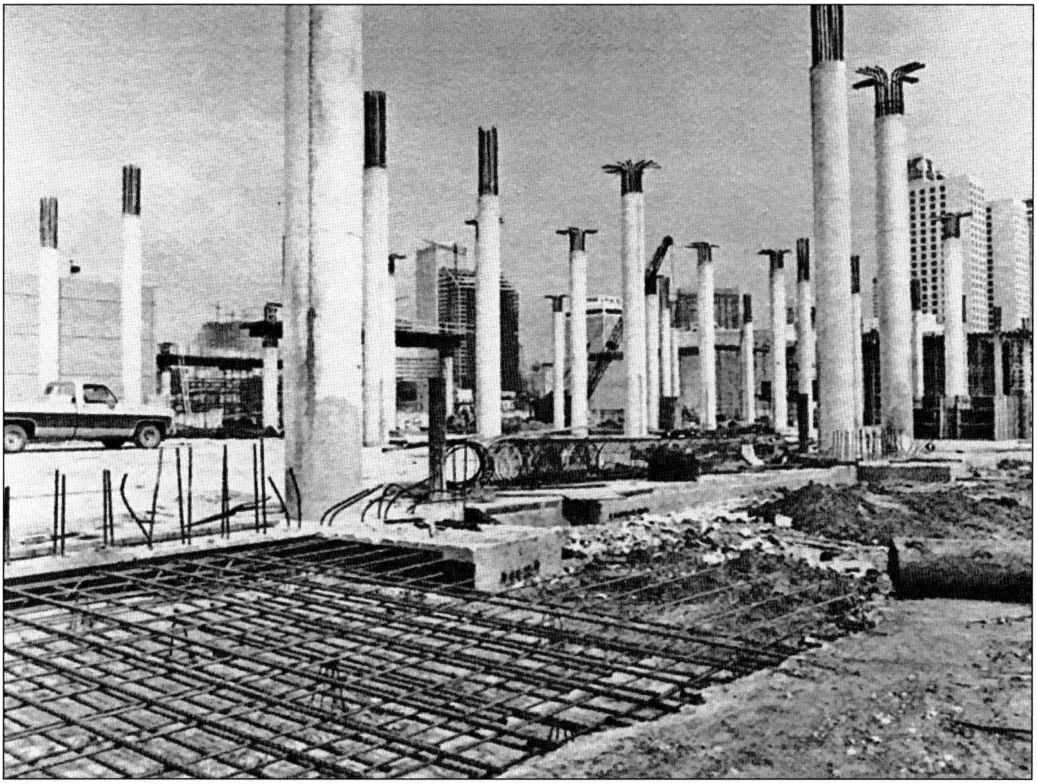

When the designs were finally done and approved, it was time to start moving dirt and bending steel. Work on the site was a complicated task with so many people needing to work in a small space at the same time. Here (above) the support pillars for the United States Pavilion take shape.

While architects and construction crews were busy planning the buildings for the fair, and other crews worked on the shows and films, there was a myriad of other details that needed attention, such as the design of the fair's logo. After several interim versions, the final logo included a nod to the "World of Rivers" theme.

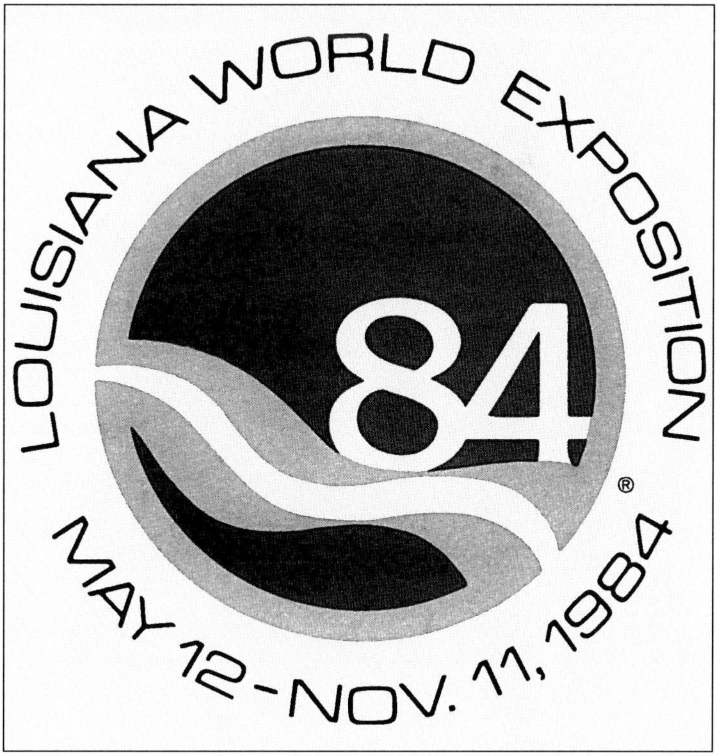

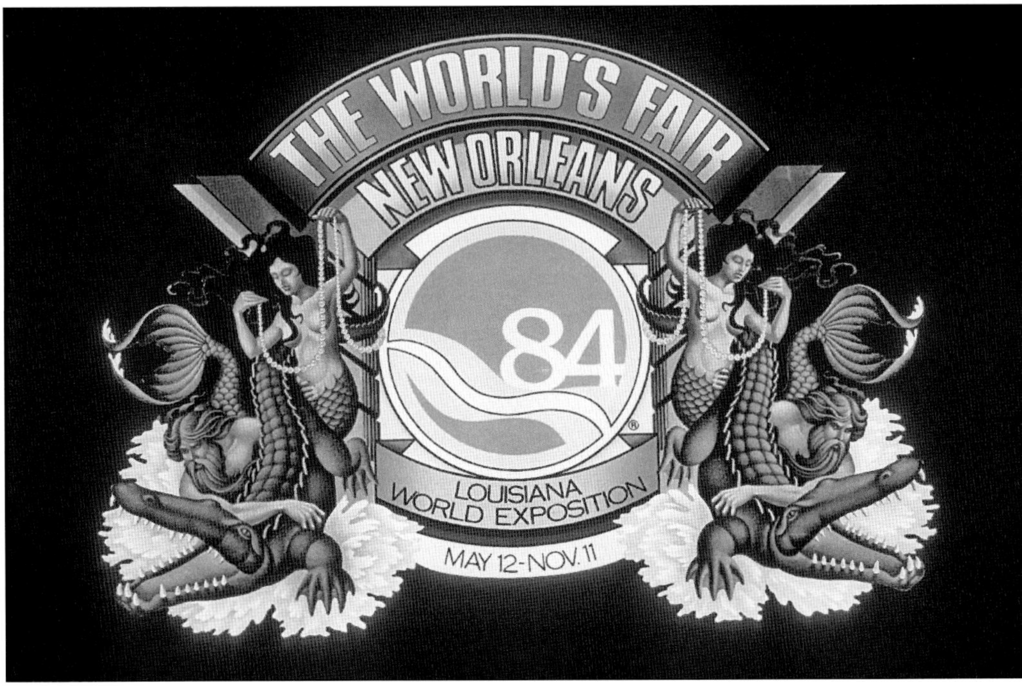

The fair also produced a second version of the logo that incorporated figures similar to those at the Bridge Gate and City Gate entrances. The simplified version of the logo was used for most souvenir items, and this one appeared on tickets, awards given to participants, and other more formal fair documents.

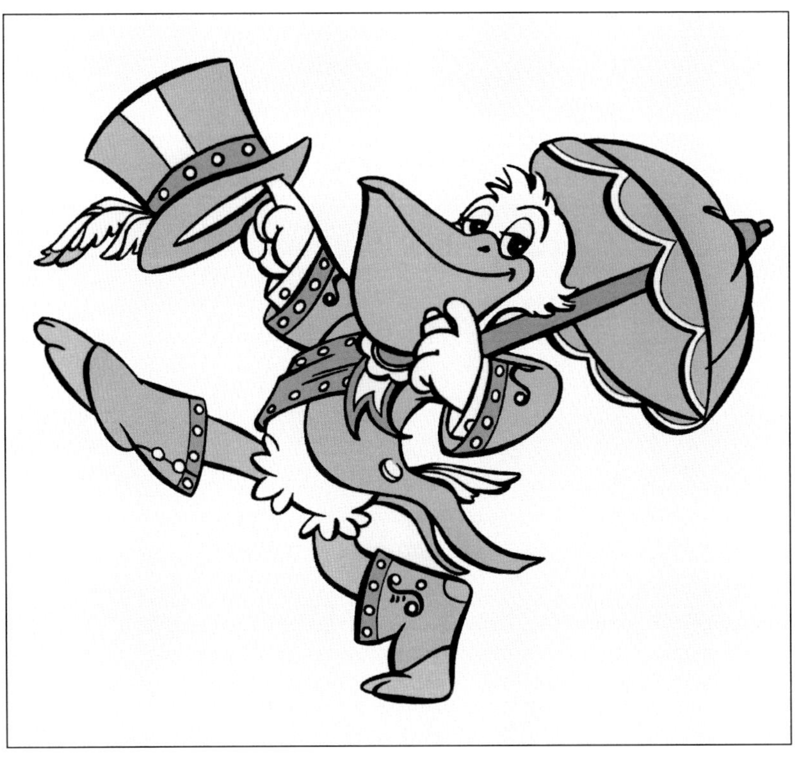

Another creation of the fair artists was the fair's mascot. Originally just referred to in press releases as "our pelican mascot" or simply "the pelican," as the fair's opening neared, he finally got a name—Seymour D. Fair.

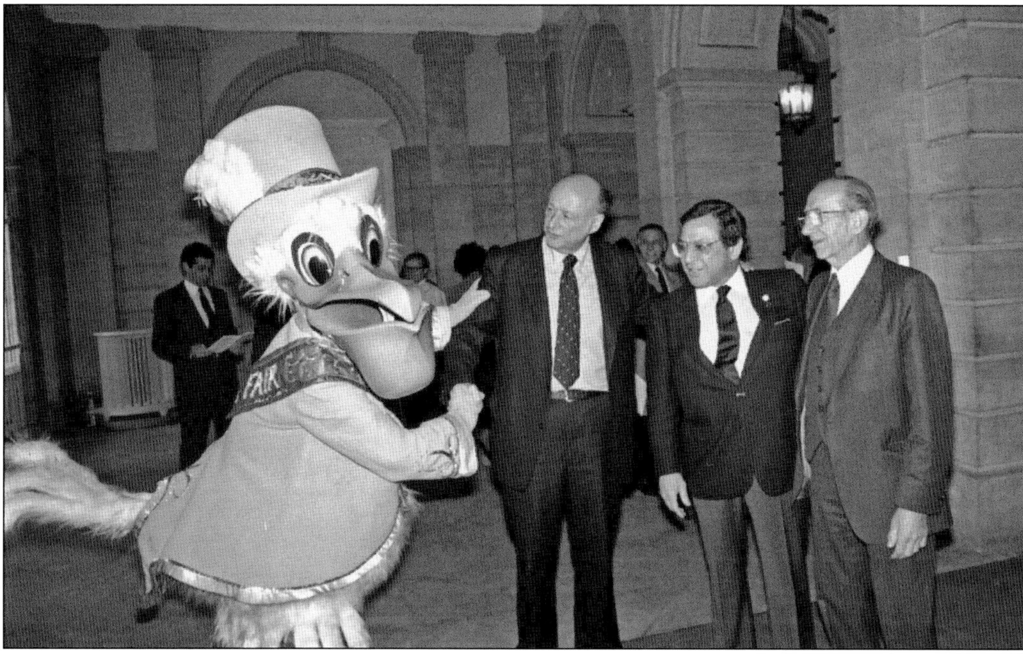

Seymour was sent on the road as part of the fair's marketing and publicity campaign. Here he is seen with Mayor Ed Koch of New York City (left) and Mayor Dutch Morial of New Orleans (center). The man on the right is unidentified. Poor Seymour suffered from an identity crisis, for some fair literature and souvenirs named him Seymore or Seemore.

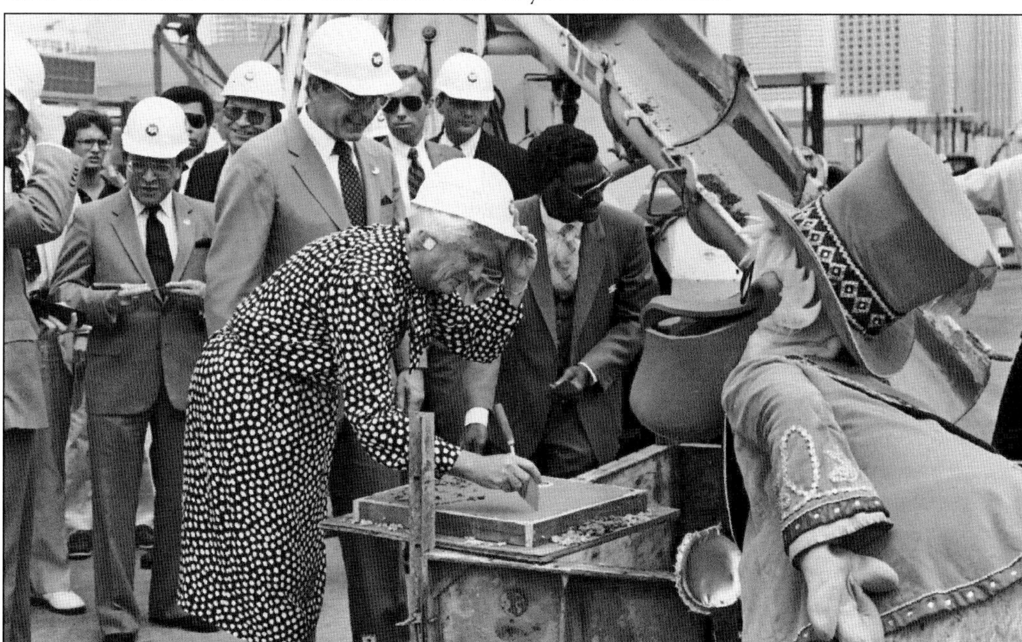

Seymour also appeared at numerous events at the fairgrounds, such as this scene of the groundbreaking ceremony for the United States Pavilion. Barbara Bush is seen putting her name in concrete, with New Orleans mayor Dutch Morial (left) and Petr Spurney (right) behind Vice Pres. George H. W. Bush.

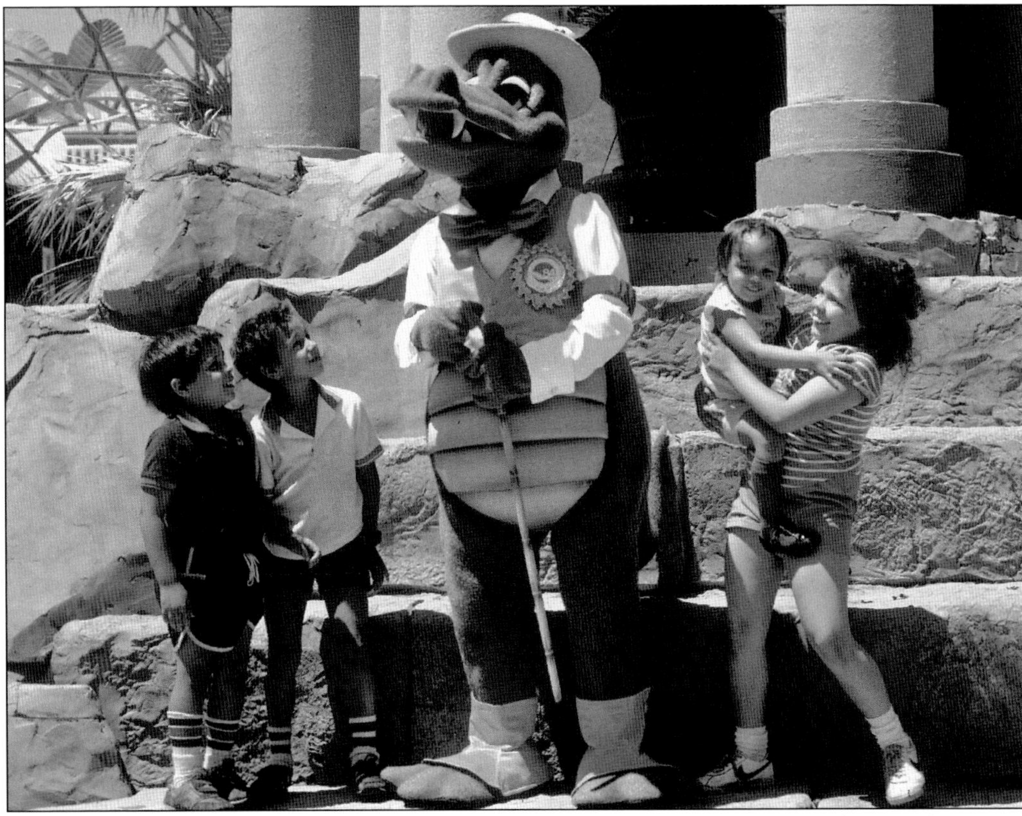

The fair also had a second mascot, an apparently unnamed alligator. Seen here welcoming a group of young visitors to the fair, he frequently appeared with Seymour in parades and stage shows. Despite that fame, the anonymous alligator was not part of the fair's merchandising efforts. (Courtesy of Mitchel Osborne.)

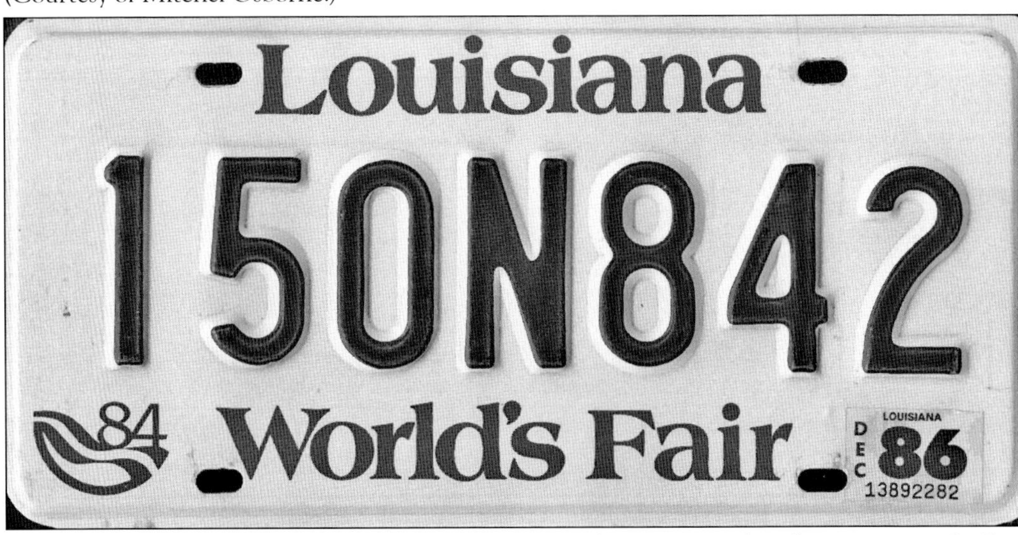

Although this was to be the New Orleans World's Fair, the state joined in the promotional efforts by adding the logo and name to all license plates issued that year. With renewals, these plates could be seen for years after the fair had closed.

Two

Centennial Plaza

Located just inside the City Gate, Centennial Plaza is a tribute to the 1884 New Orleans Cotton Exposition and a time when Cotton was King and New Orleans was the Queen City of the South.

—1984 Annual Report

When the designers decided to break the fair into different themed areas to deal with the physical challenges posed by the site, the largest open area became known as Centennial Plaza. Designed to capture the look and feel of the 1884 Cotton Exposition, the central point of the plaza was a fountain-filled pool known as Centennial Lagoon. The lagoon introduced visitors to the fair's theme of "The World of Rivers: Fresh Water as a Source of Life," with a lively show featuring computer-controlled fountains matched to several different music tracks.

While the area was named and themed after an event held 100 years earlier, Centennial Plaza was also the home of several exhibits that showcased the present and the future. One of Louisiana's major industries at the time of the fair was offshore oil drilling, and visitors could explore a full-size oil platform just like those found in the Gulf of Mexico. They could also learn about the nation's electrical system and how it impacted their lives, or those in a more adventurous mood could travel several hundred feet above the Mississippi River on a new mass-transit gondola system.

While other parts of the fair were built with plans to reuse them after the fair ended, no such plans were formulated for Centennial Plaza, and all of the buildings were removed after the fair closed. The area today is a parking lot for the Riverwalk Mall and the Convention Center, and the only visible remnants of the fair at street level are a pair of bridges to the former International Riverfront area. When viewed from overheard though, one can still see the ghostly outlines of Centennial Lagoon and many of the buildings that once welcomed visitors to the fair.

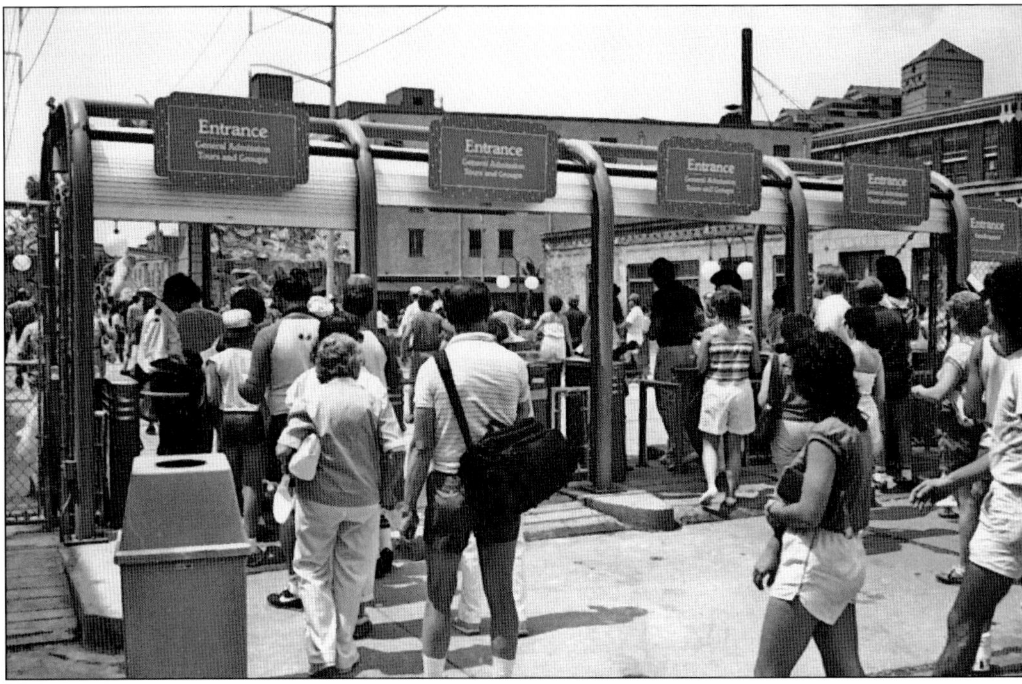

Centennial Plaza was one of the two main entry points into the fair. While most guests arriving by car would have entered through the Bridge Gate, those living nearby or traveling via public transportation would have found themselves at these ticket booths in front of the City Gate entrance.

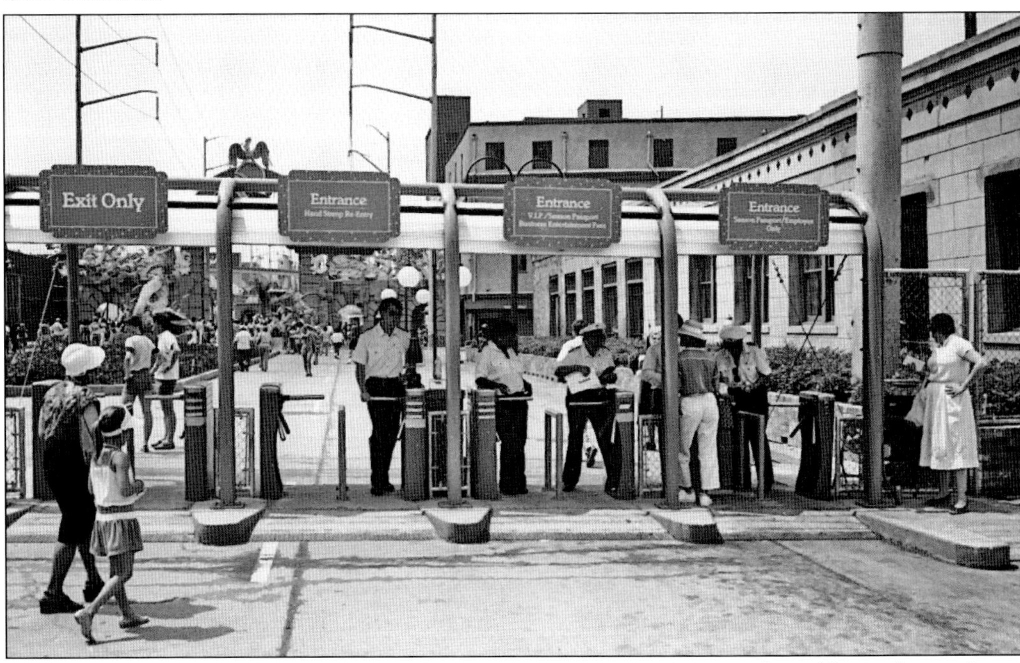

The fair's temporary nature can be seen in the traffic lane markings still visible on the surface of Front Street, which led the way towards the City Gate. The crush of last-minute construction to get the fair open did not include time for details like removing the stripes.

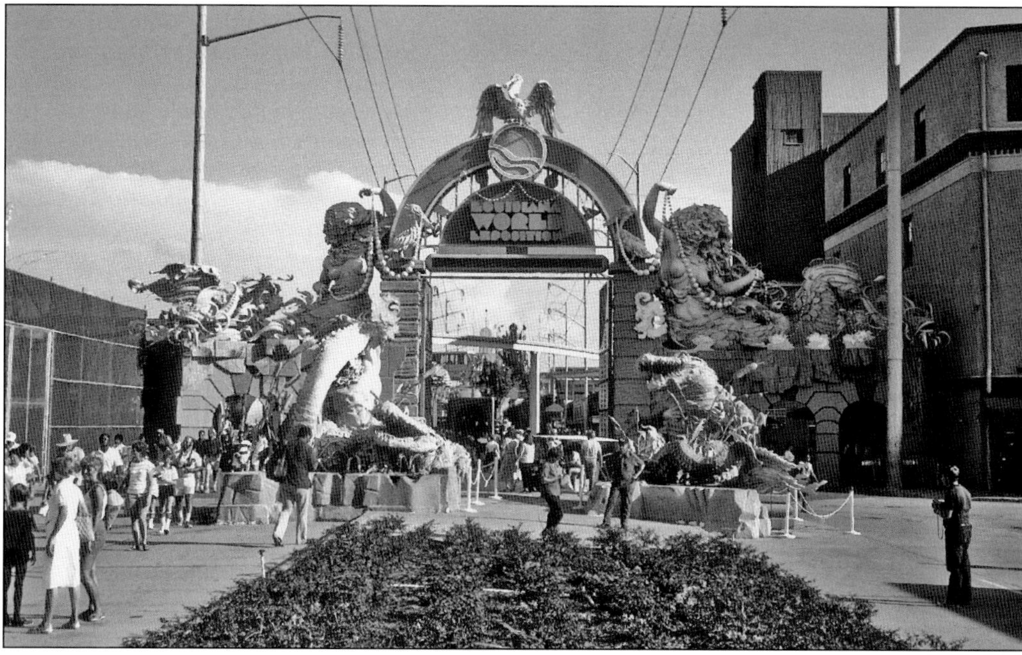

The first view most visitors had of the fair itself was the fanciful City Gate. The 51-foot-high gate is perhaps the most remembered image of the fair and was featured in much of the fair's promotional material. The power lines on either side show how the site challenged the designers.

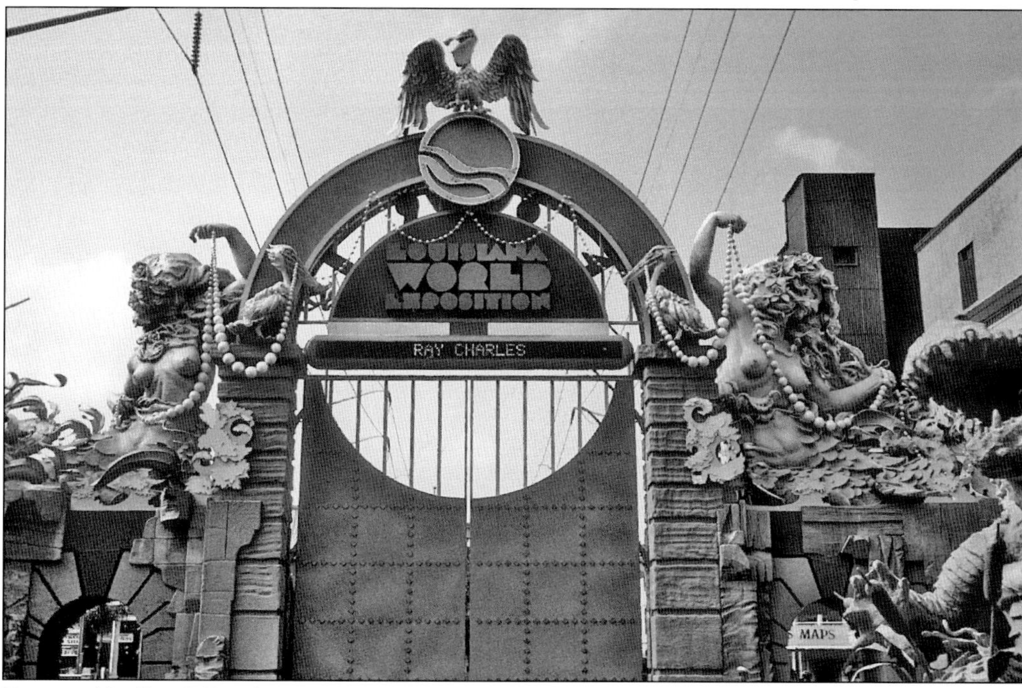

Designed by Barth Brothers, Inc., a prominent Mardi Gras float-building company, the City Gate generated a great deal of commentary—and controversy—due to the mostly naked mermaids on either side of the main entrance. The figures undoubtedly helped sell a great deal of extra film to visiting tourists who just had to share this experience with the folks back home.

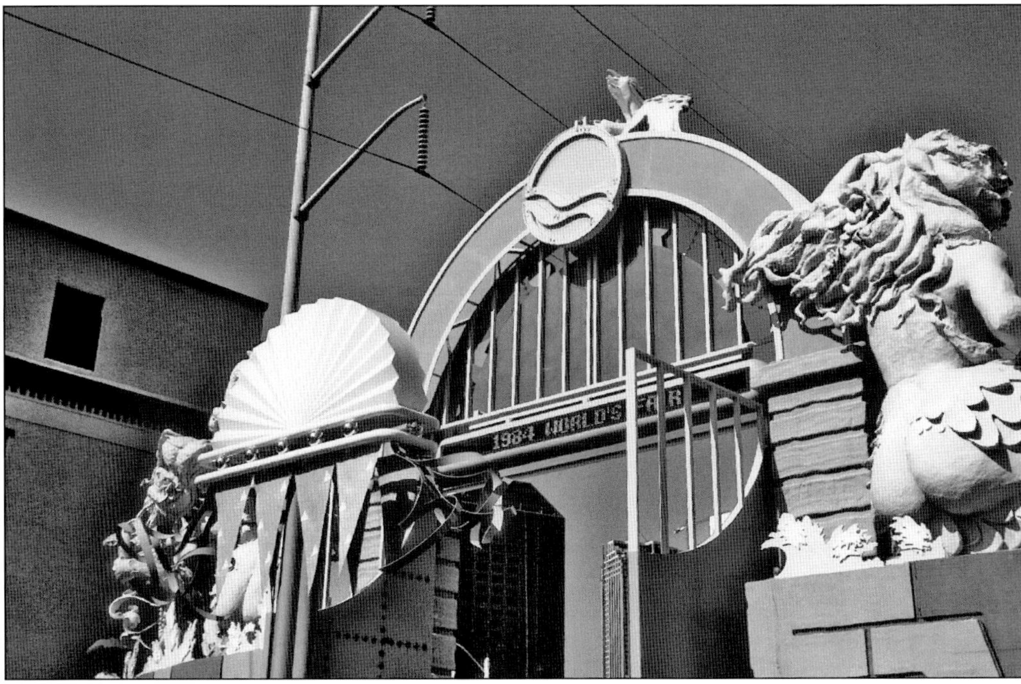

The original colors of the figures were painted over to give them a more classical look, but the controversy continued throughout the life of the fair, much to the amusement of those who participated in the far more rowdy Mardi Gras celebrations each year.

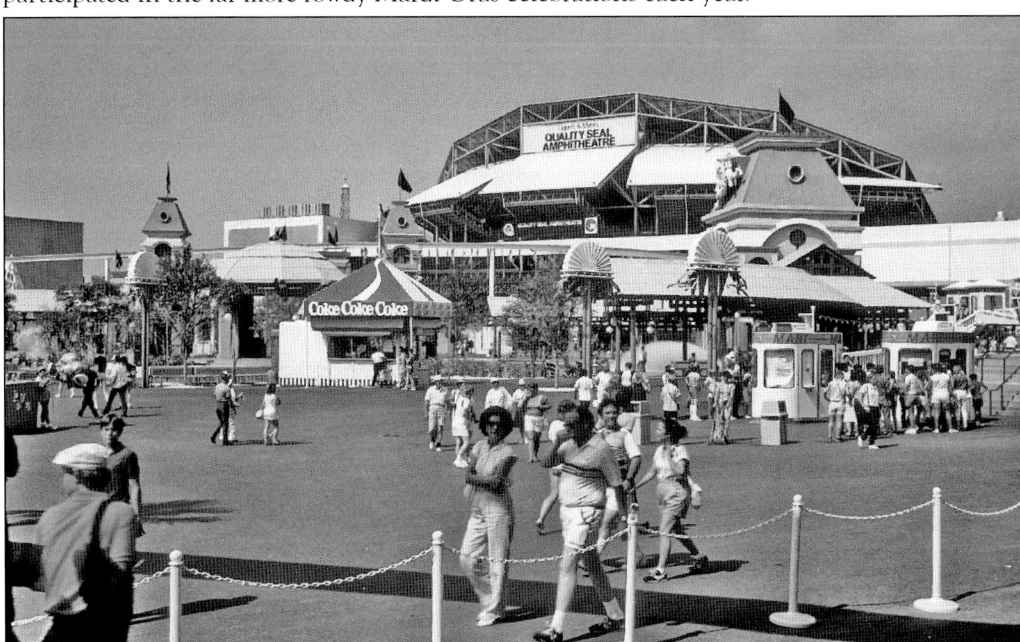

Inside the gate was Centennial Plaza, a large open area flanked by several pavilions and restaurants that surrounded the Centennial Lagoon. Many of the buildings were patterned after parts of the 1884 World's Industrial and Cotton Centennial, providing a link to the city's first major international exposition.

One of the most visible structures in the area was the Petroleum Industries Pavilion, a full-size oil-drilling platform. Visitors could climb through the structure for a lesson on the importance of offshore drilling, one of Louisiana's major industries.

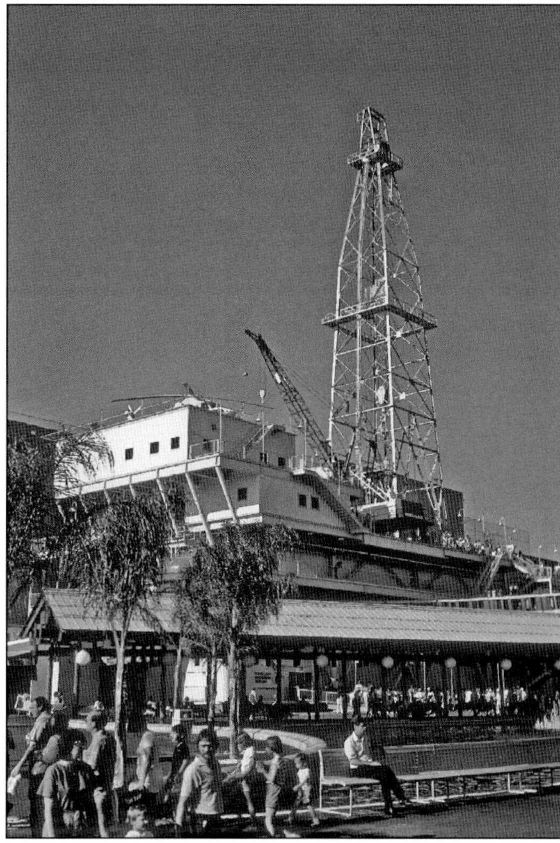

Inside the 20-story pavilion, guests were treated to a 50,000-gallon salt-water aquarium followed by a dual-screen 35-millimeter film titled *Reflections*, which explained the pavilion's theme, "Where Oil and Water Meet." Actual drilling tools, including a 12-passenger helicopter, were also on display.

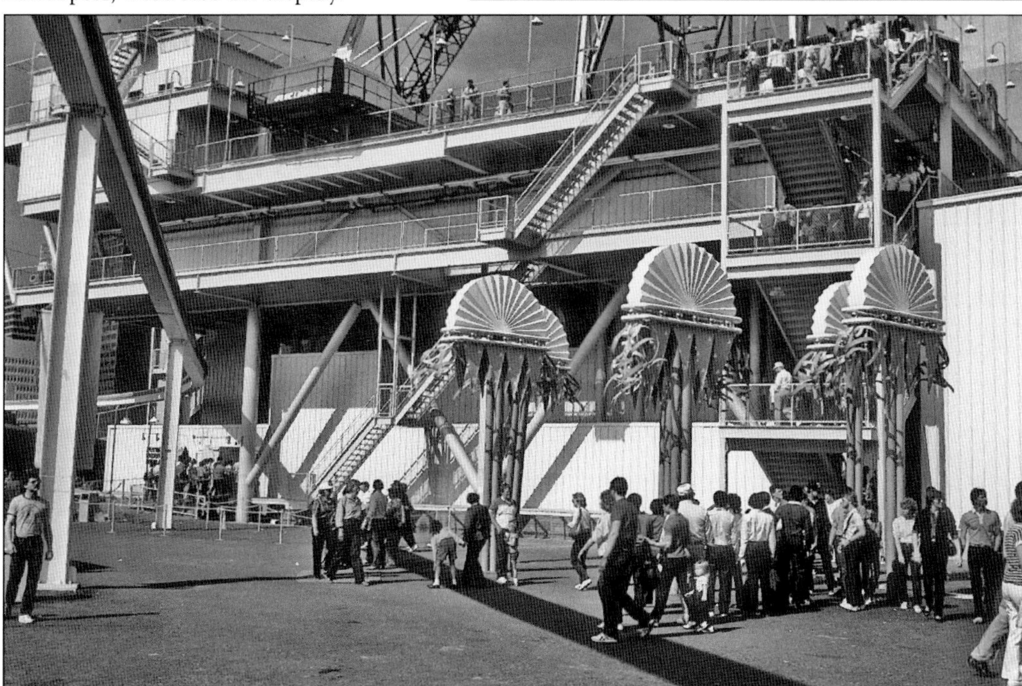

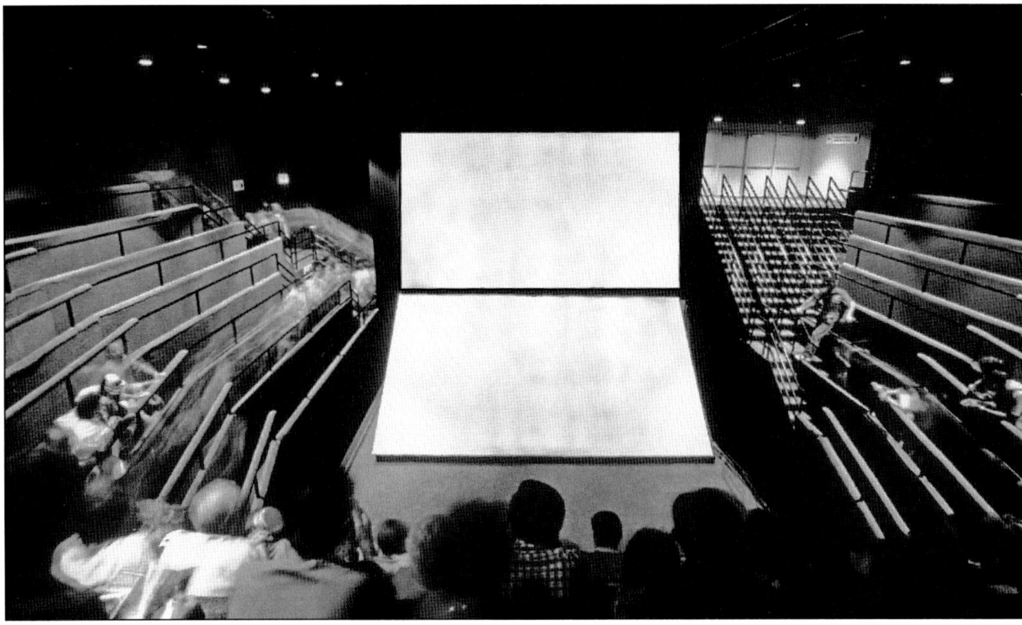

The pavilion's film was shown on an unusual screen. While other dual-screen films had been shown side-by-side, the Petroleum Industries Pavilion's *Reflections* used a horizontal set of screens, which allowed the filmmakers to show life on the oil platform both above and below the sea. (Courtesy of Barry Howard.)

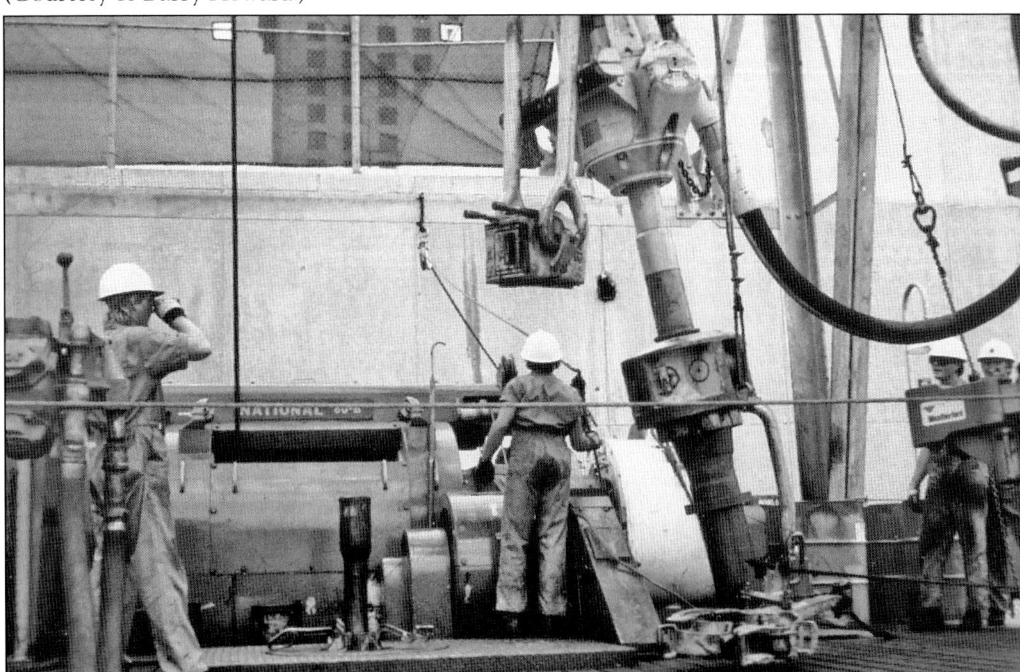

Many residents of Louisiana were only familiar with oil platforms from a distance, but at the world's fair, they could see one up close. The pavilion included ongoing demonstrations of the techniques used in oil exploration, demonstrating the tools used on the real platforms out in the Gulf of Mexico. (Courtesy of Barry Howard.)

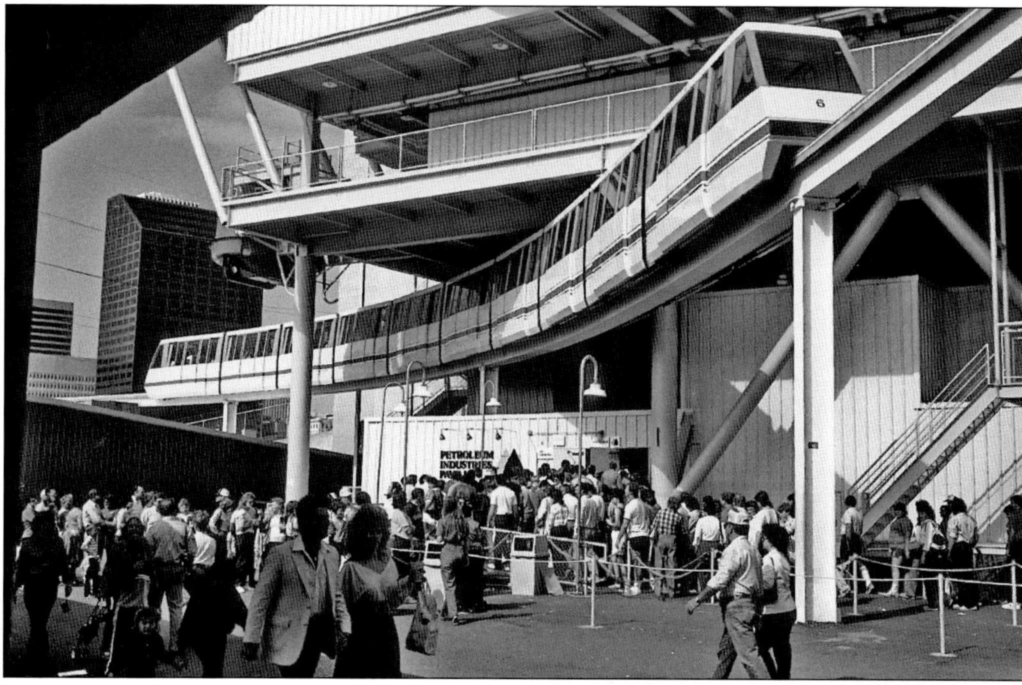

This view of the monorail passing the Petroleum Industries Pavilion illustrates one feature of the fair that garnered a great deal of criticism. There was plenty of asphalt and steel, but little in the way of trees or other greenery to help cool the site down.

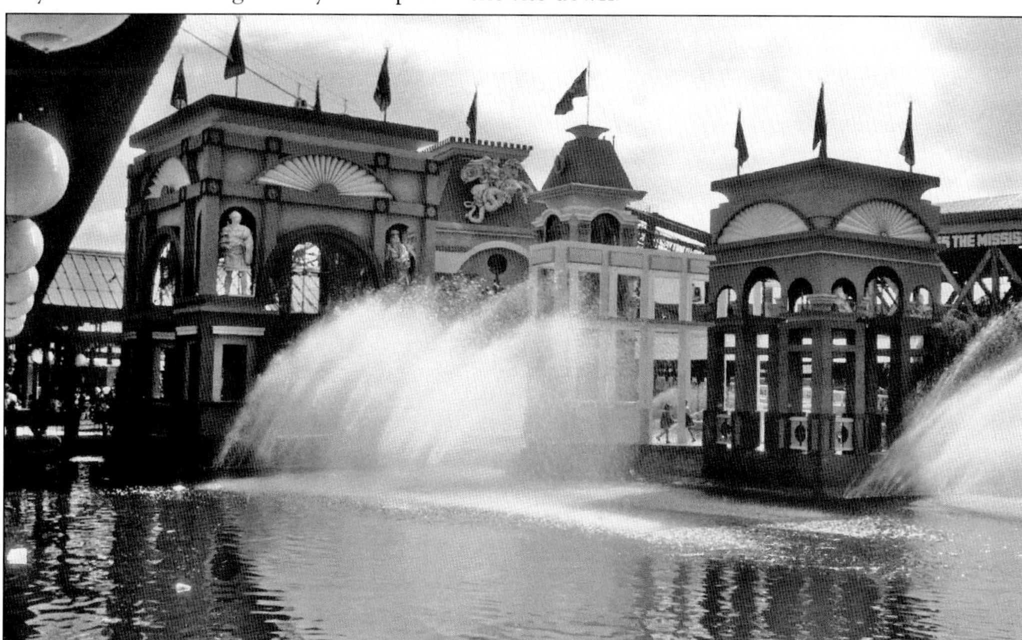

Located on one end of the Centennial Lagoon, these structures were loosely based on pavilions that had been built for the 1884 Cotton Exposition. The seven buildings, known as the Centennial Pavilion, were basically giant gazebos without any displays or exhibits, but at least they did provide some shade.

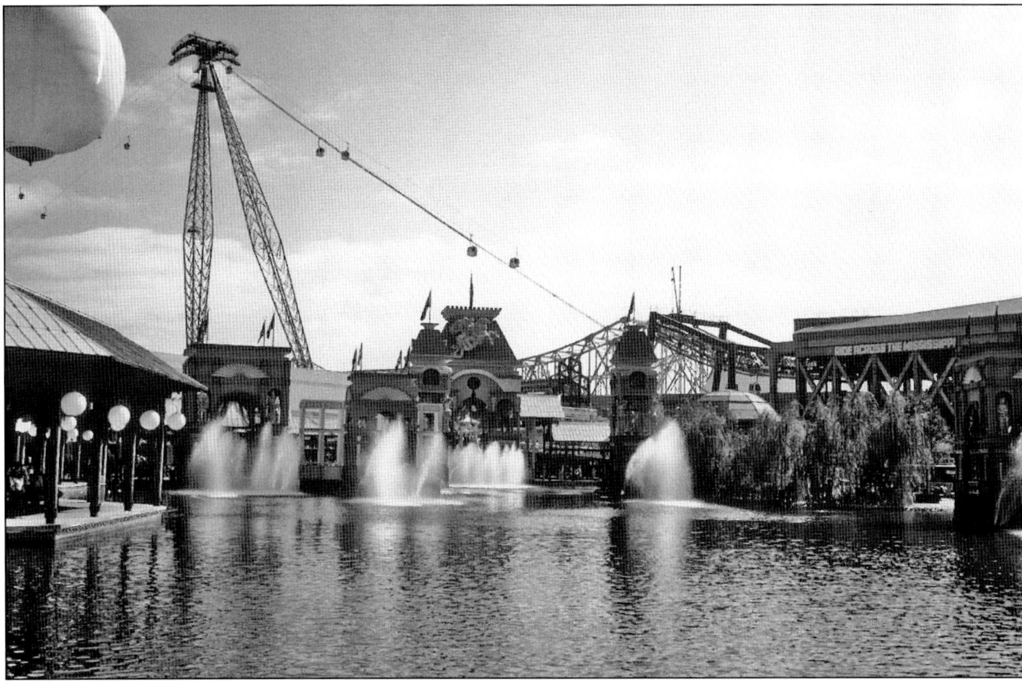

The fountains of Centennial Lagoon leap into the air in front of the Centennial Pavilion, with the MART gondola system rising into the sky in the background. The fountains could be programmed to entertain visitors with several different shows set to a variety of music.

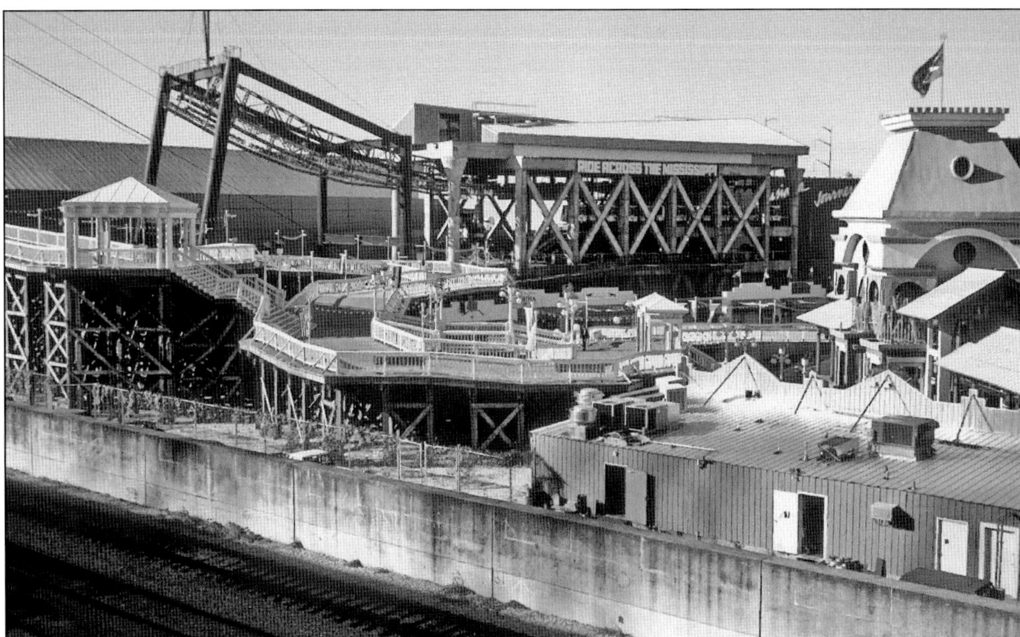

MART stood for "Mississippi Aerial River Transit." Fairgoers could board a gondola for a trip high across the river. Like the smaller Sky Transpo in the International Riverfront section, MART was based on a system designed for ski resorts. This is the station on the fair side. A similar one was in the Algiers section of the West Bank across the river.

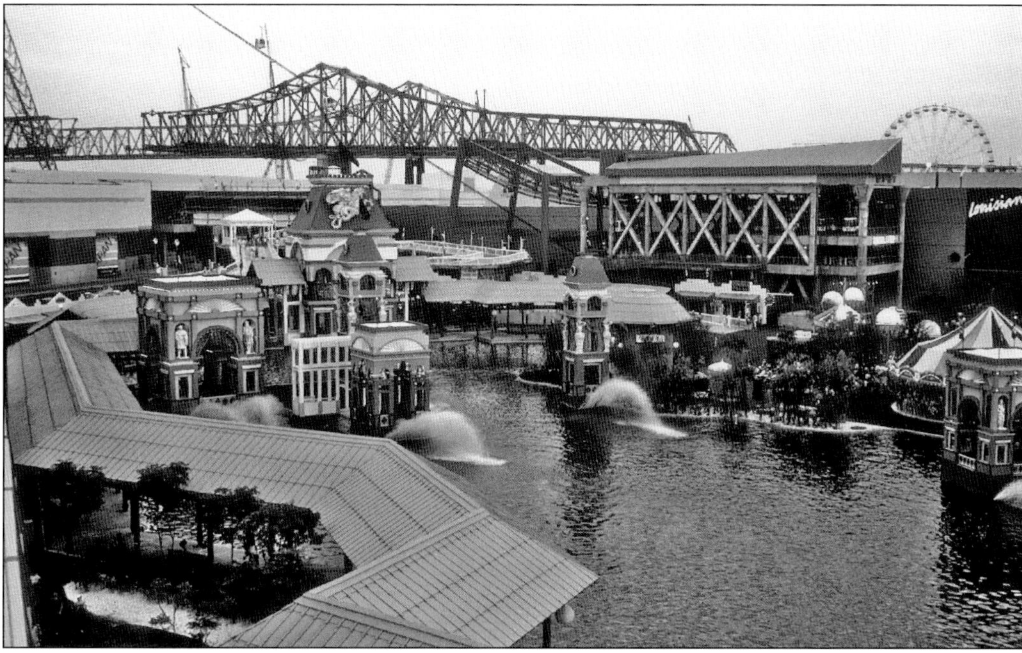

The cables of the $8-million MART system rise steeply into the sky behind the fountains and facades of Centennial Plaza. The lift towers were the tallest ever built for a gondola system, standing 358 feet high. The roof of the Empress Walk is seen on the left, with the Centennial Pavilion at the end of the lagoon.

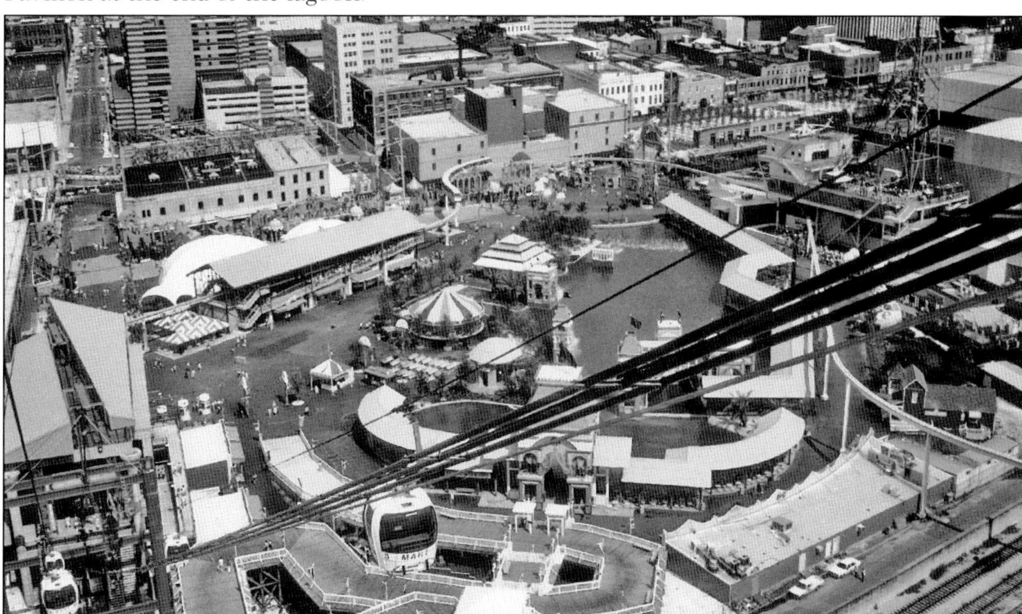

MART provided a great way to see all of Centennial Plaza, as shown in this view from a car descending into the station. The waters of Centennial Lagoon are in the center of the photograph, and the Empress Walk crosses underneath the MART cables off to the right. The domes of America's Electrical Energy Exhibit can be seen poking above the roof of the monorail station to the left of the lagoon.

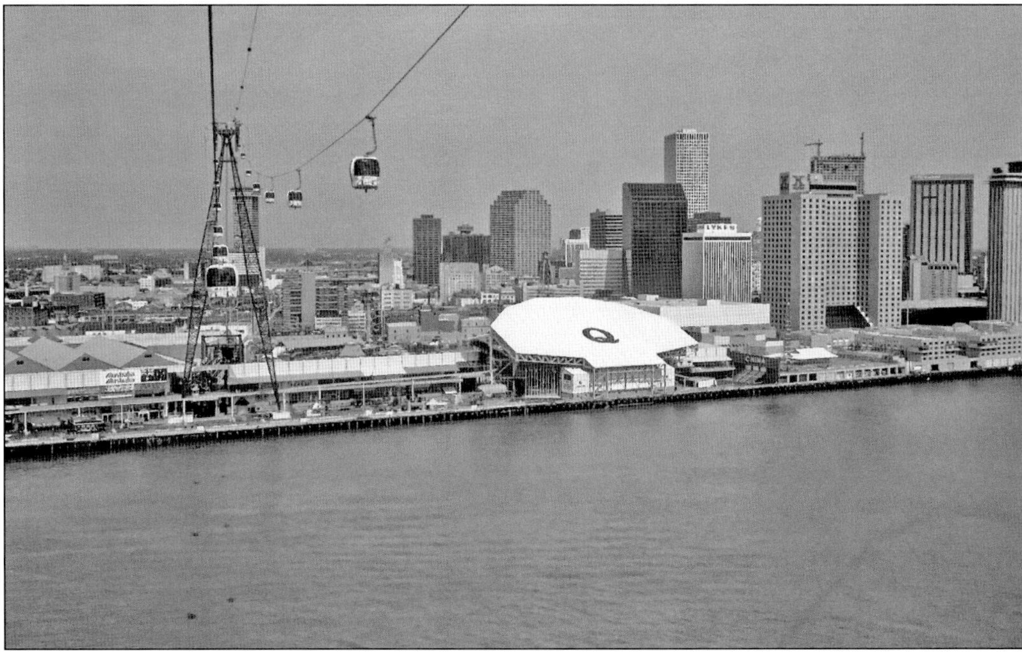

MART rose 300 feet above the Mississippi River, carrying up to 2,000 passengers per hour on the four-minute ride. The system was intended to remain in use after the fair, and when the Tower of New Orleans project fell apart, fair officials tried to compare MART to other world's fair legacies such as the Eiffel Tower and the Space Needle.

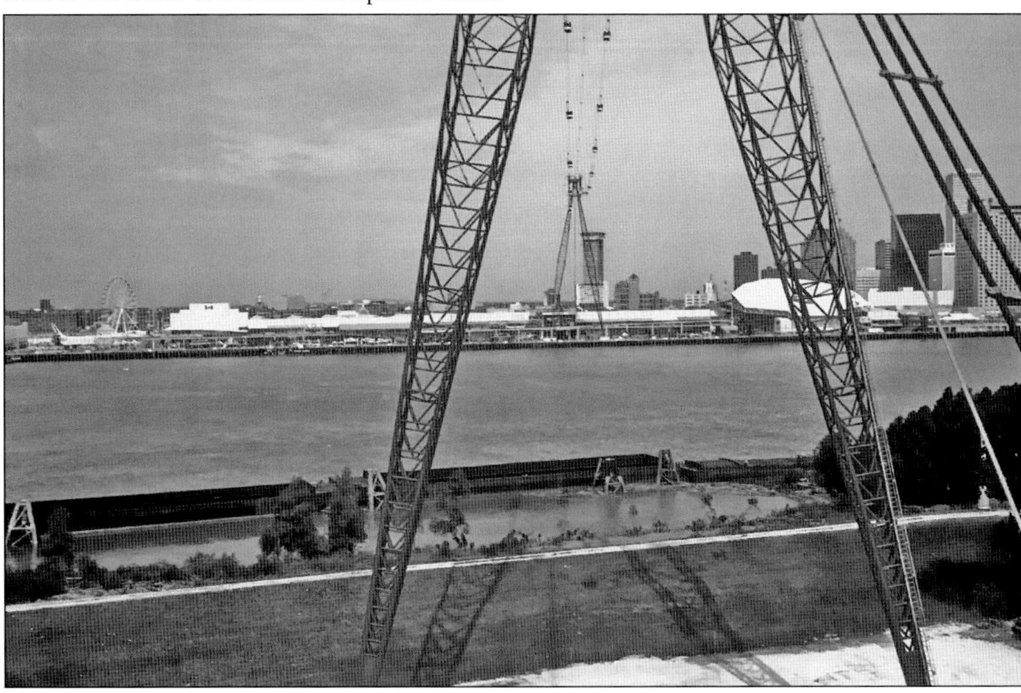

This view from the Algiers station shows just how far away the fair looked from the other side of the river. Many fairgoers found the ride unsettling, especially on windy days, but once across the river, they had few other options to get back to the fair. The round-trip experience cost $3.50.

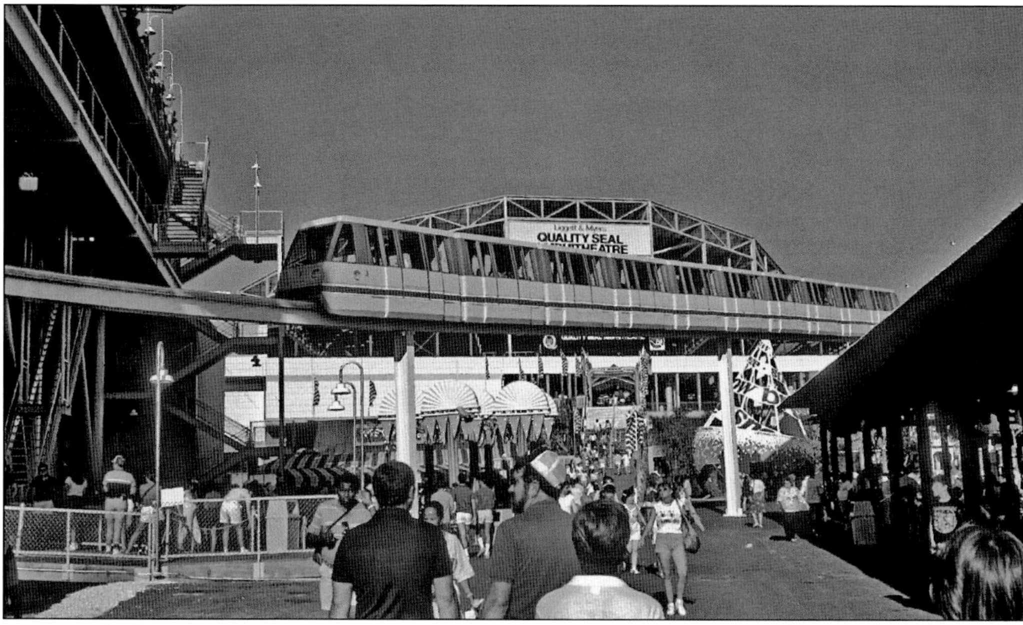

Fairgoers looking for an easier ride experience could enjoy the monorail, seen here near Centennial Plaza. The 146-passenger trains circled the site and stopped at Centennial Plaza, Bayou Plaza, and the International Riverfront. The roof of the Quality Seal Amphitheatre can be seen in the back, just above the monorail.

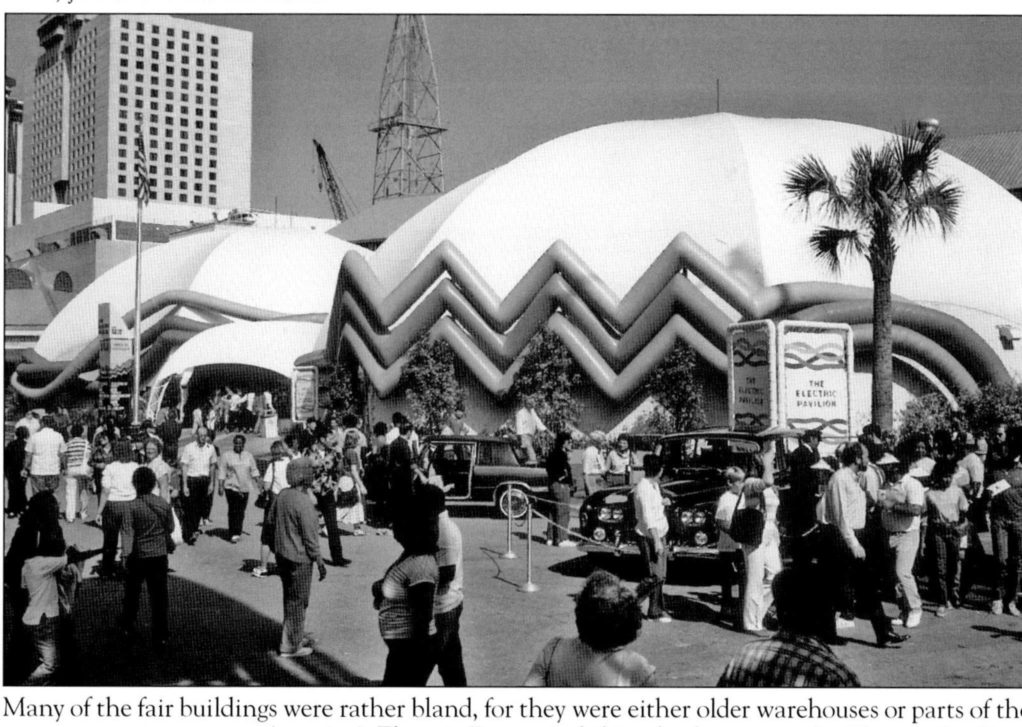

Many of the fair buildings were rather bland, for they were either older warehouses or parts of the new convention center. America's Electric Energy Exhibit, also known as the Electric Pavilion, helped make the area more interesting with its two unusually shaped domes, which had air-filled tubes in contrasting shapes and colors.

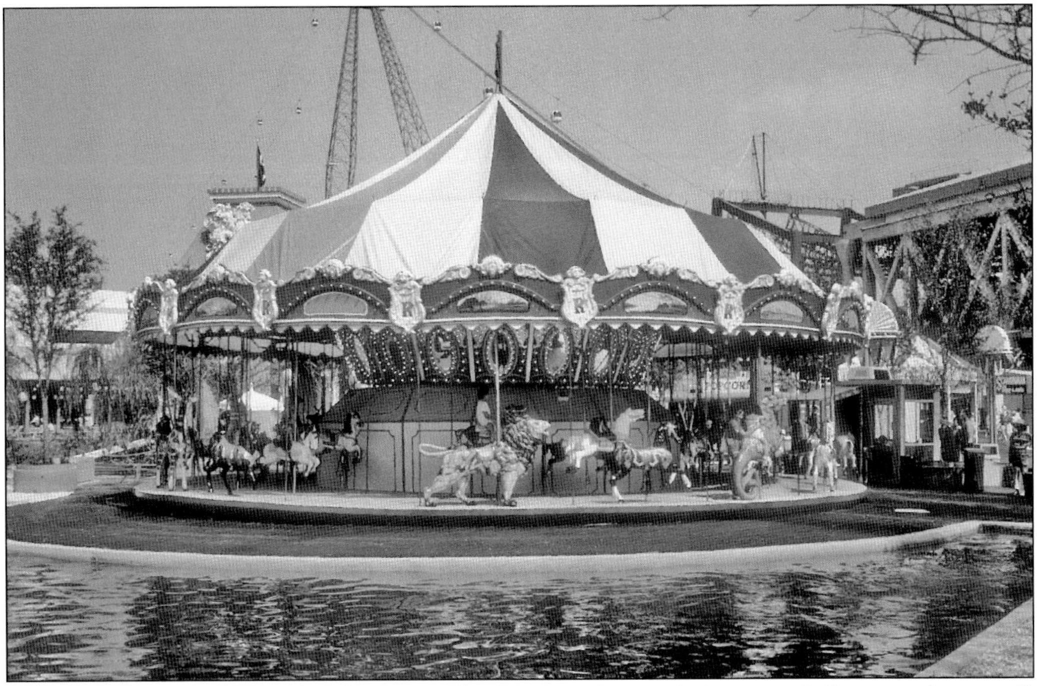

Set up at the edge of Centennial Lagoon, the Antique Carousel also provided some color as well as fun for children of all ages. Built in 1904, the ride was popular with younger visitors, for most of the fair was decidedly intended for older guests.

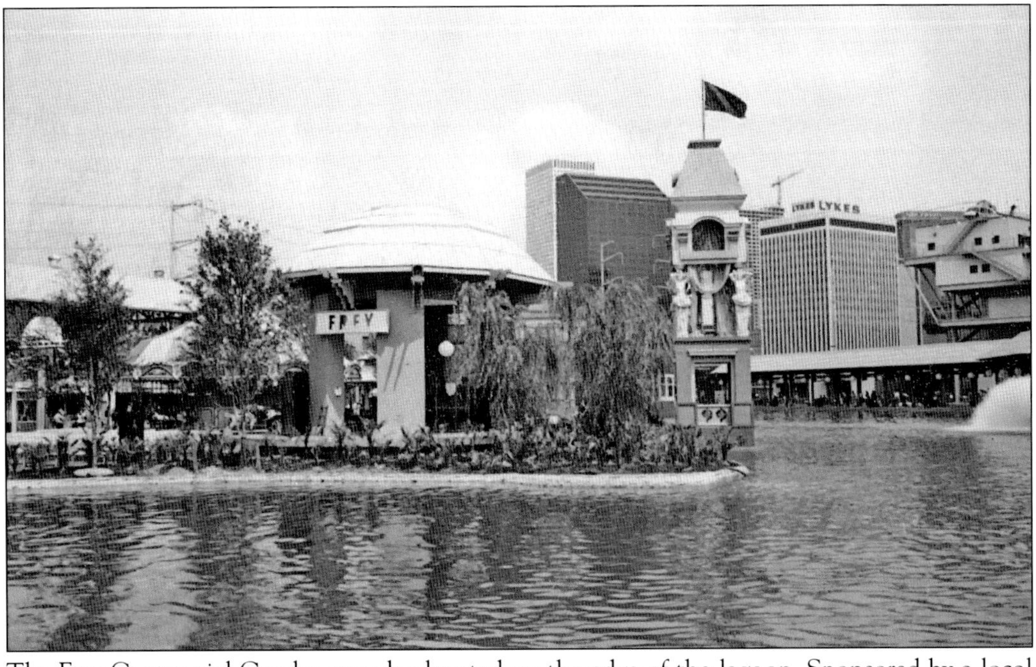

The Frey Centennial Gazebo was also located on the edge of the lagoon. Sponsored by a local meatpacking company, the 200-seat gazebo hosted performers with early-20th-century themes, such as barbershop quartets and banjo players, in keeping with the overall look and feel of the Centennial Plaza area.

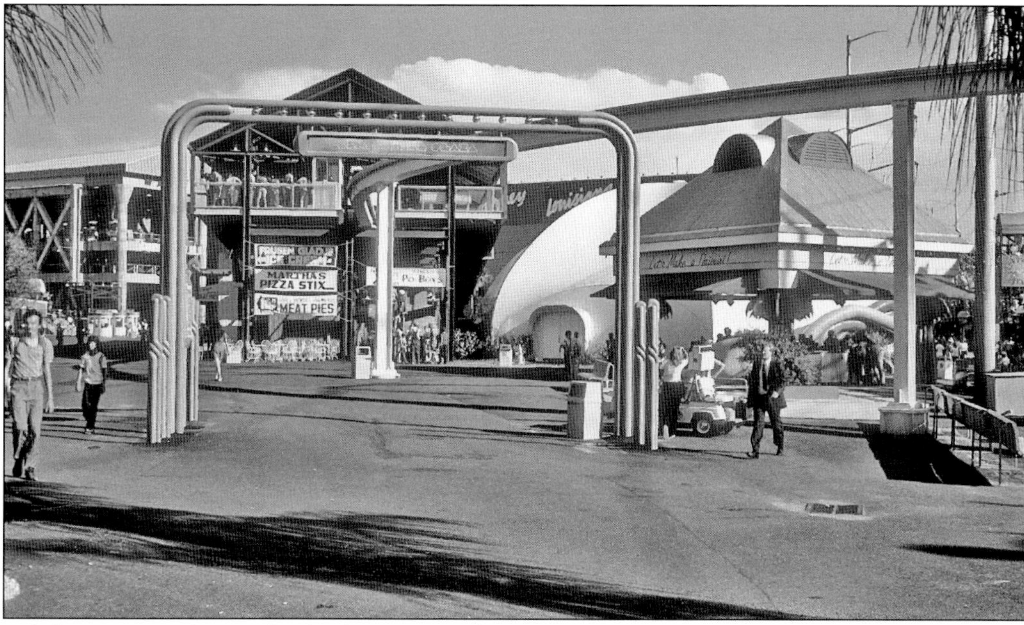

While many vendors at the fair lost money due to the smaller-than-expected crowds, the owners of the Let's Make a Daiquiri! concession found the fair to be a very profitable venture. Their shop on the right of the picture may be empty in this view captured early in the morning, but by the time the usual New Orleans heat arrived, the place would be thronged with thirsty guests ready for a cold drink.

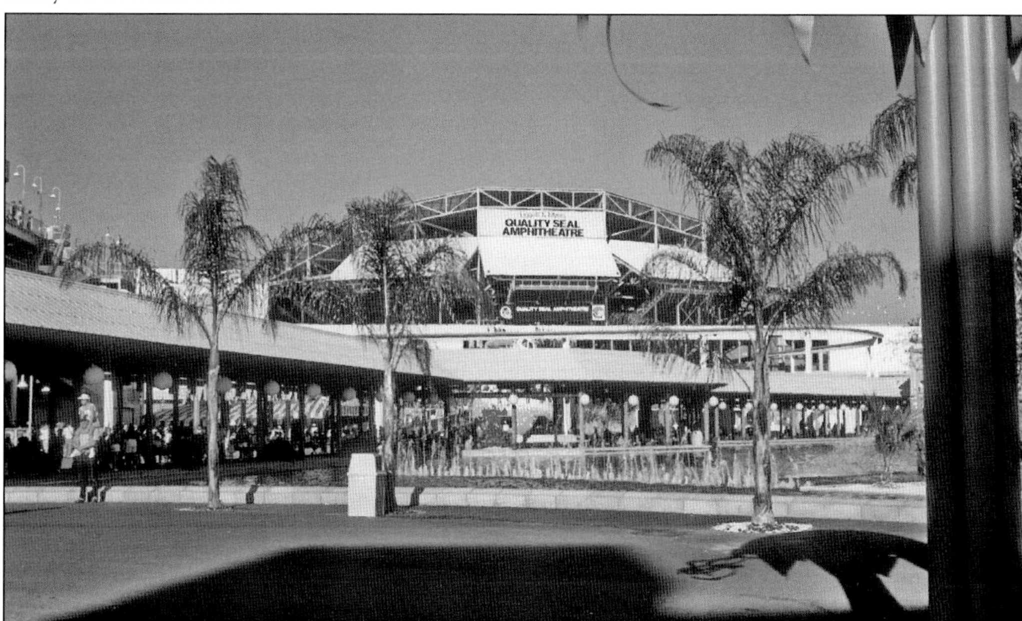

Supposedly inspired by the Empress's Summer Palace in Peking, the Empress Walk was described as a place of peace and calm, offering visitors a tranquil spot to view activities in the area. Stretching along most of the lagoon, it is best remembered as another way to get out of the sun as guests crossed the plaza to the rest of the fair.

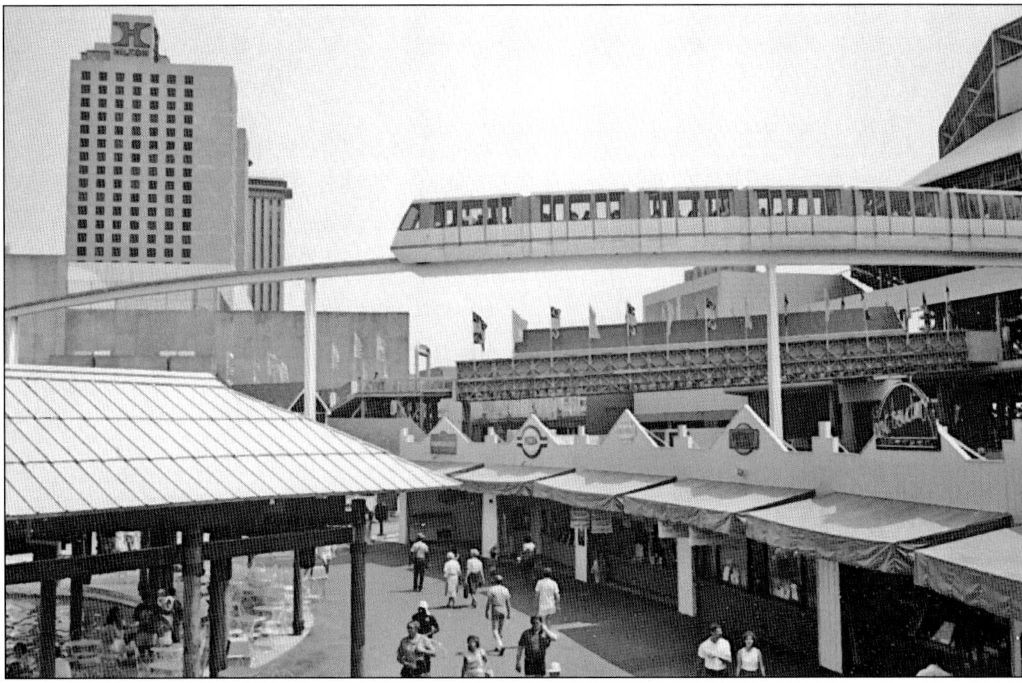

Just across from the Empress Walk was a wide assortment of food shops. The most popular of these was Petro's, which served up a tasty mix of sour cream, chili, tomatoes, and onions, all poured into a bag of Fritos corn chips. This unlikely combination was the food hit of the fair.

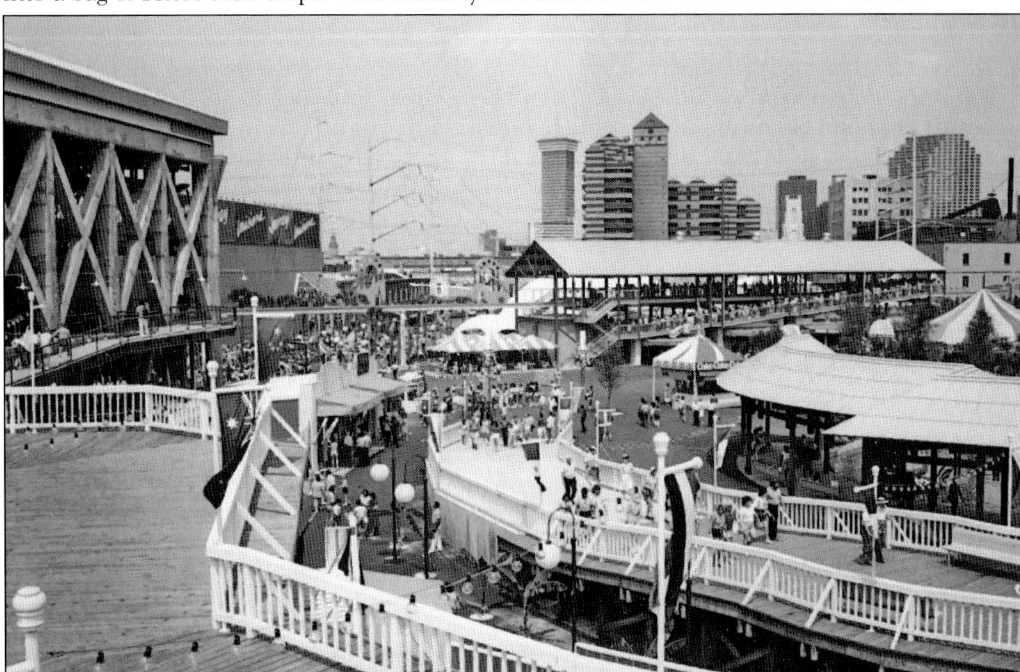

After enjoying the plaza area, visitors could travel to the International Riverfront area via the Centennial Ramp. Flags of the participating nations on each side set the mood for the exhibits to come and provided a splash of color to the oversized boardwalk.

Three

Fulton Street Mall

Lined with historic, restored warehouses, Fulton Street is an inviting, spacious pedestrian mall celebrating New Orleans as a Mecca for international shopping, food, and music.

—1984 Annual Report

World's fairs have always been known for pushing the envelope in their architectural designs, usually challenging designers to create structures that stretch the imagination. How does one build a fair though, when a large part of the site consists of run-down old buildings that cannot be removed or even significantly altered? That was exactly the problem encountered at what would become the Fulton Street Mall, which was originally to be named Gaslight Square. To make matters worse, the buildings were owned by numerous companies and individuals, and the fair had to negotiate a separate lease agreement with each of them.

The decision was made not to build the mall around any major exhibition, but instead to develop it as a collection of shops, restaurants, and nightclubs that would build upon local themes. Reunion Hall, for example, was a 17,000-square-foot venue that featured live jazz and gospel music and headlined famous clarinet player Pete Fountain. The surrounding buildings were filled with smaller restaurants, bars, and gift shops.

The only new development in the area was the very popular Vatican Pavilion, operated by the Archdiocese of New Orleans. The 1964–1965 New York World's Fair's Vatican Pavilion had been a major success, due in large part to the exhibition of Michelangelo's famous statue, *Pietà*. For this fair, the Vatican loaned an impressive number of paintings, sculptures, and religious artifacts, making this one of the must-see attractions of the fair.

The fact that the original buildings that became the Fulton Street Mall could not be torn down turned out to be a major benefit for New Orleans, for the area now hosts a thriving art community, with trendy restaurants and shops mixed in with galleries and expensive lofts. Just as the fair backers had planned, the fair did indeed act to rejuvenate a significant part of the city that once seemed to have no future.

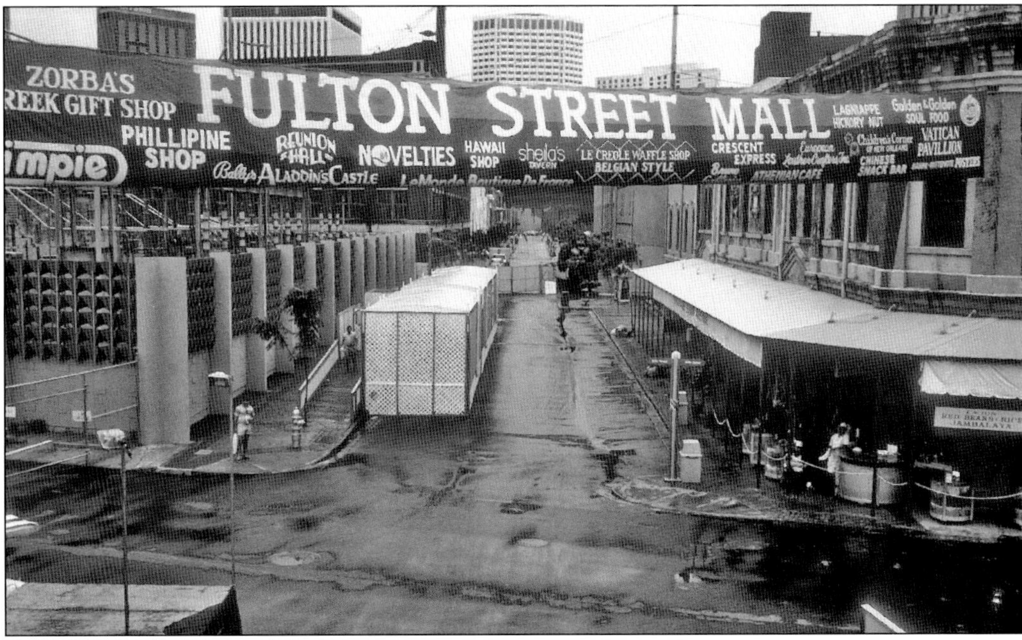

The Fulton Street Mall certainly didn't look like a typical world's fair setting. The original industrial nature of the site can easily be seen here, for the awnings and flags didn't do much to brighten it up. While the area did look somewhat better at night, the designers had few options due to preservation laws.

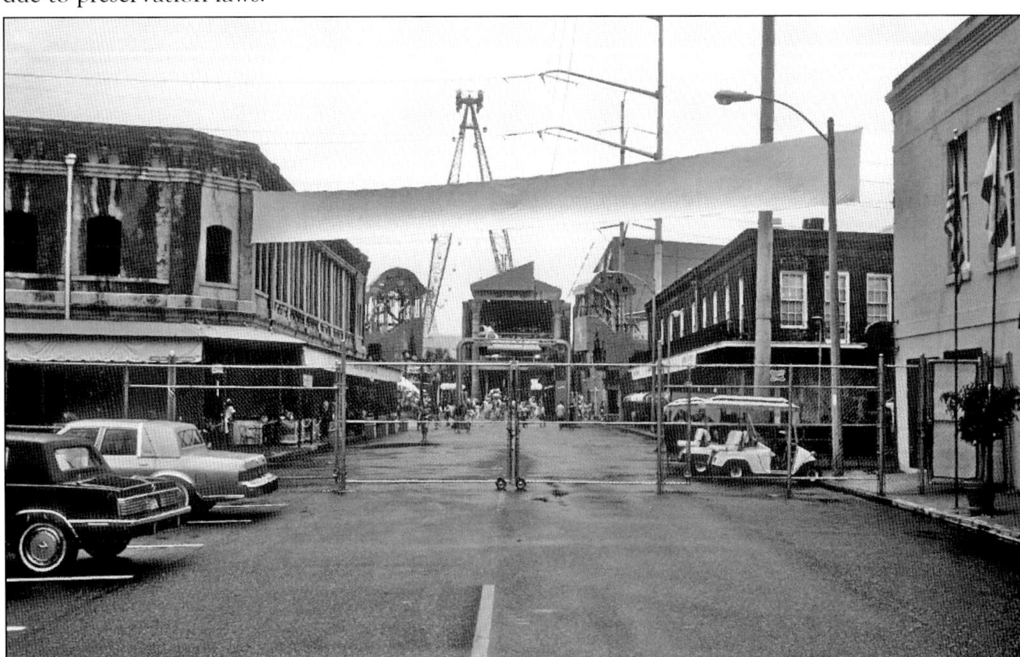

Taken from outside the fairgrounds, this view shows how a simple chain-link fence was all that separated the fair's mall area from the surrounding warehouses and other businesses. While visitors were enjoying the sights, sounds, and tastes of the fair on one side of the fence, trucks were busy hauling goods to and from factories still operating in the warehouses just outside.

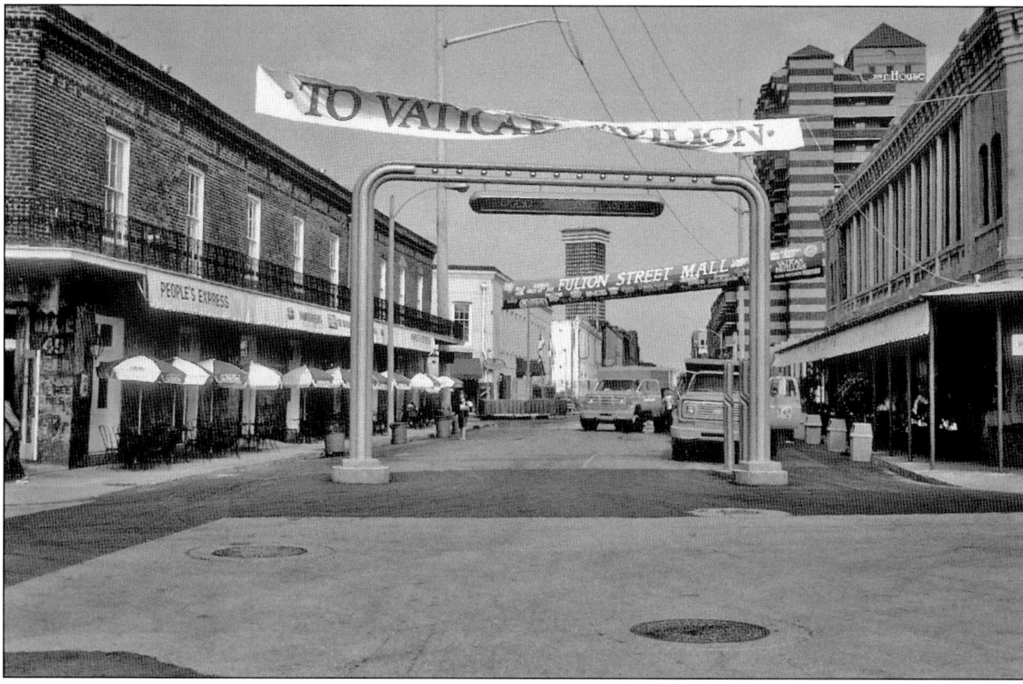

The older buildings of the Fulton Street Mall stood in marked contrast to the newer structures erected for the rest of the fair. New signage and informational arches were added to make the area look livelier, but overall, this was probably the least memorable section of the fair.

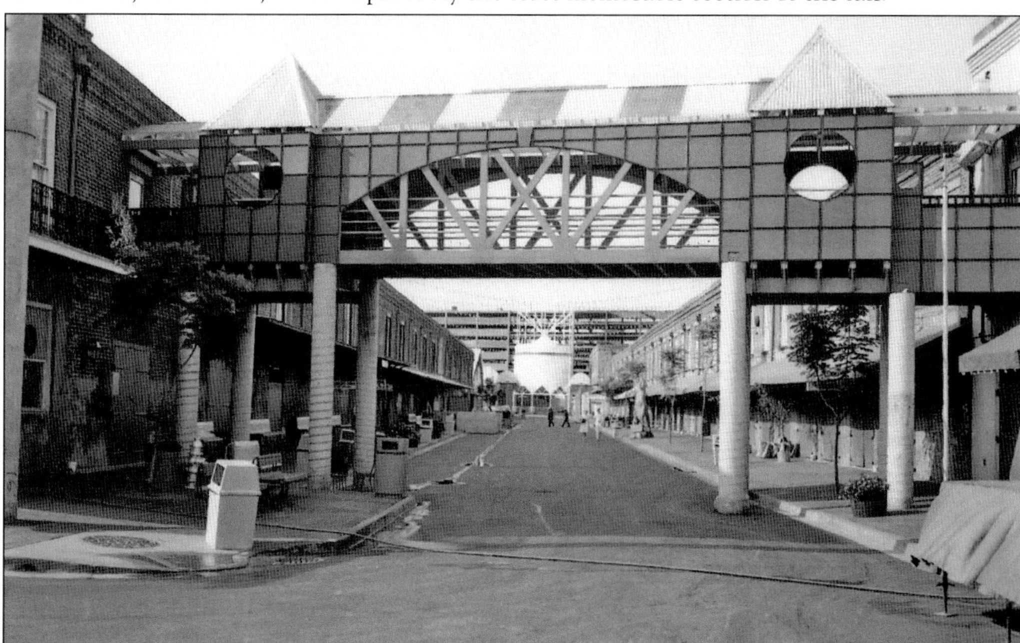

Many of the buildings required extensive repairs, including new roofs and structural strengthening, all of which had to be done at the fair's expense. This led to charges by the *Times-Picayune*, a major local newspaper, that the fair was being used to unjustly profit landowners and fair officials. None of the charges were ever proven.

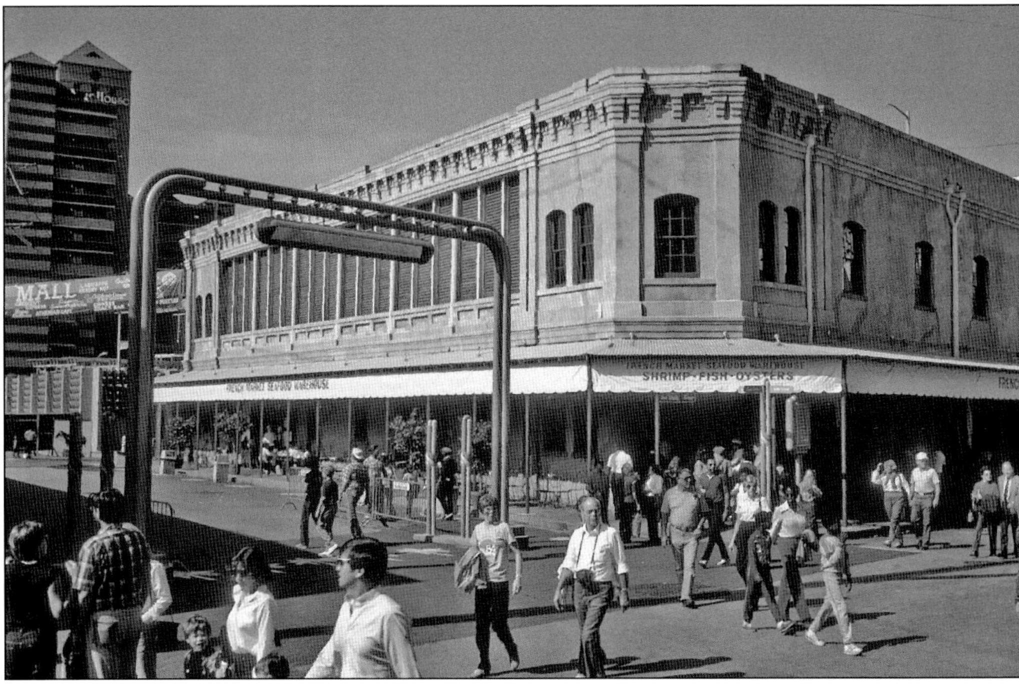

The French Market Seafood Restaurant was just one of the entertainment venues in the mall. While often quiet during the day, the area was more popular at night, when the heat dropped and the music started. Just two years before the fair, this building was in desperate need of repair.

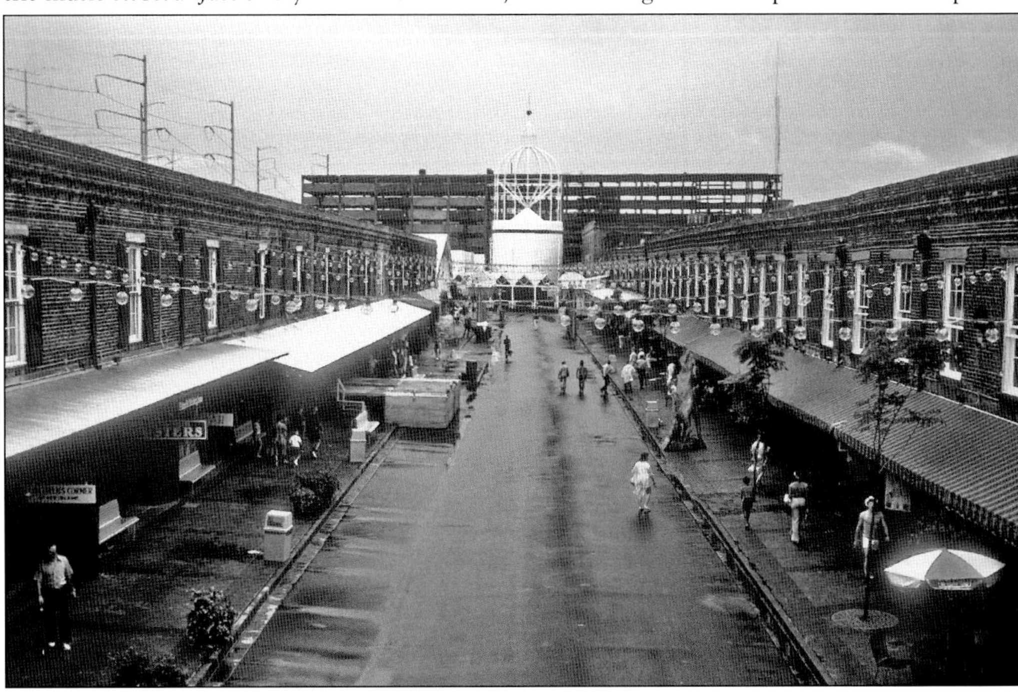

The stark white Vatican Pavilion was the only part of the Fulton Street Mall constructed for the fair. The modernistic design was a jarring architectural departure from the rest of the area, making the shortcomings of the older buildings even more noticeable.

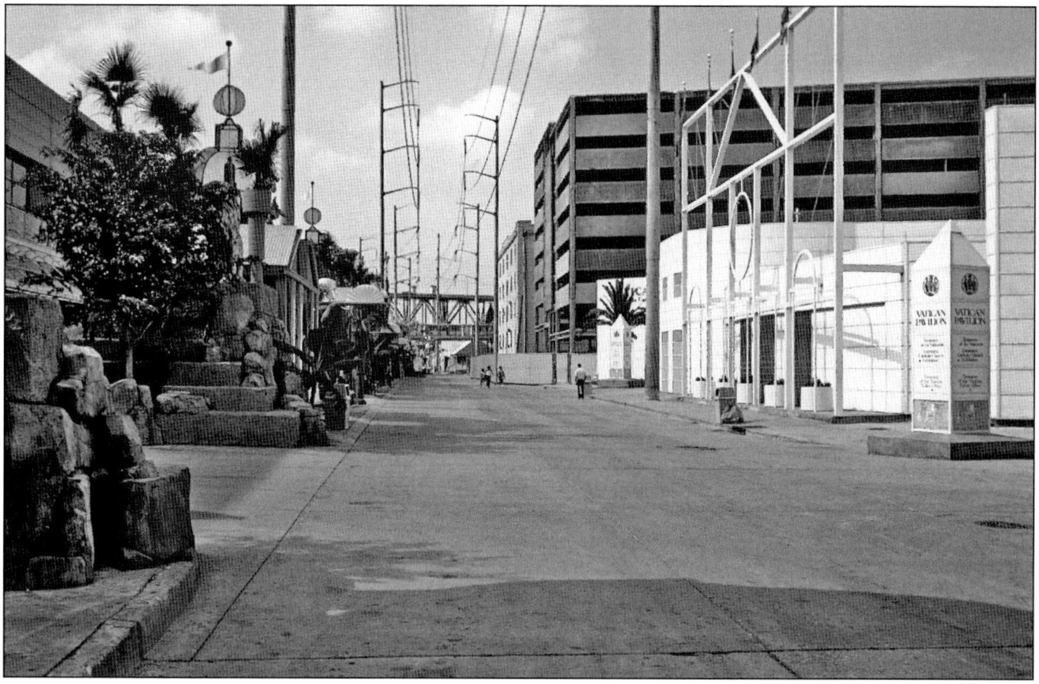

There wasn't much that could be done to camouflage this city street, which bisected the mall area. The Vatican Pavilion filled a notch in the site plan, but the office building behind it shows how the reality of the city intruded on the fantasy setting common to most world's fairs.

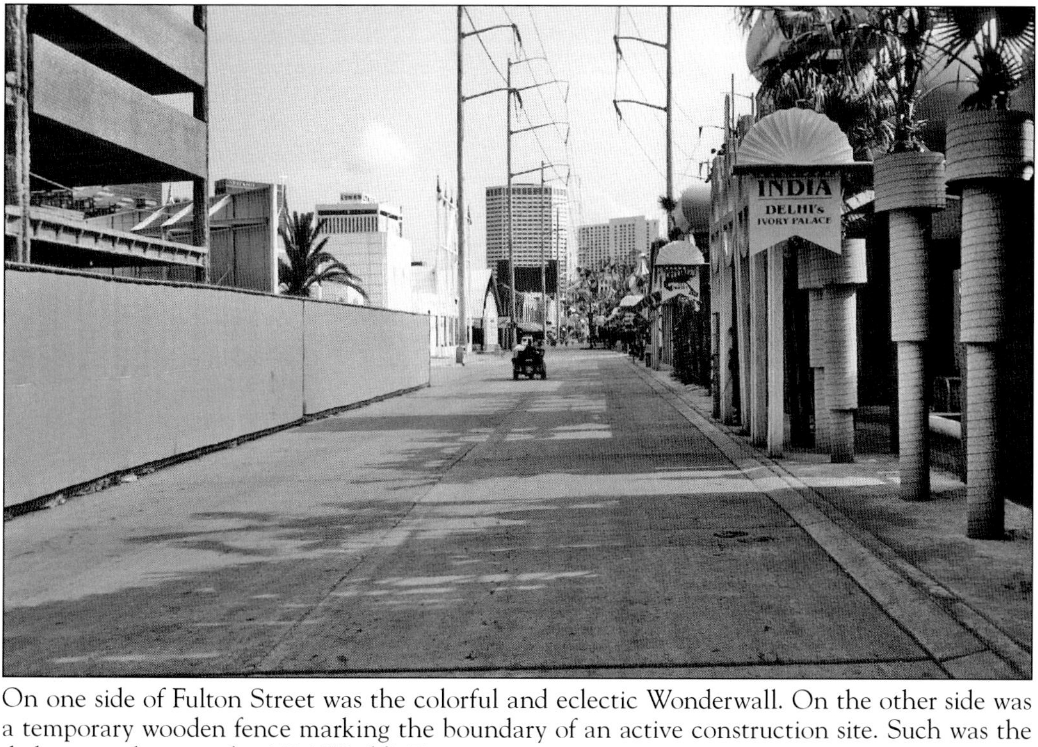

On one side of Fulton Street was the colorful and eclectic Wonderwall. On the other side was a temporary wooden fence marking the boundary of an active construction site. Such was the dichotomy that was the 1984 World's Fair.

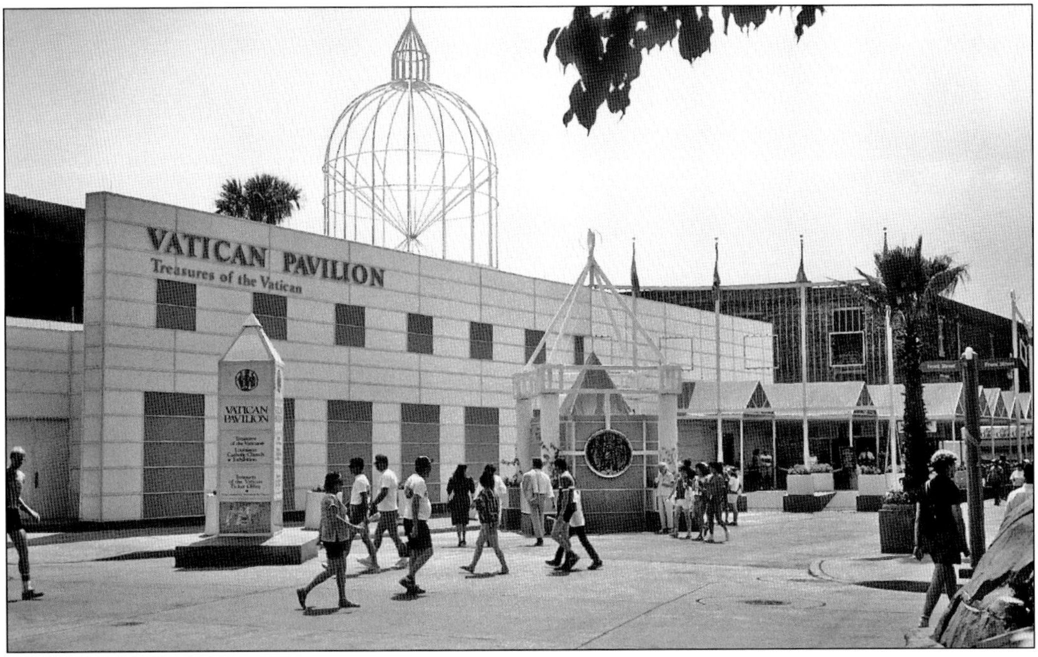

The Vatican Pavilion was unique in that it charged $5 for admission, the only pavilion with a separate entrance fee. Despite the fee, it was one of the most popular attractions at the fair, and art scholars and journalists praised the number and quality of the exhibits.

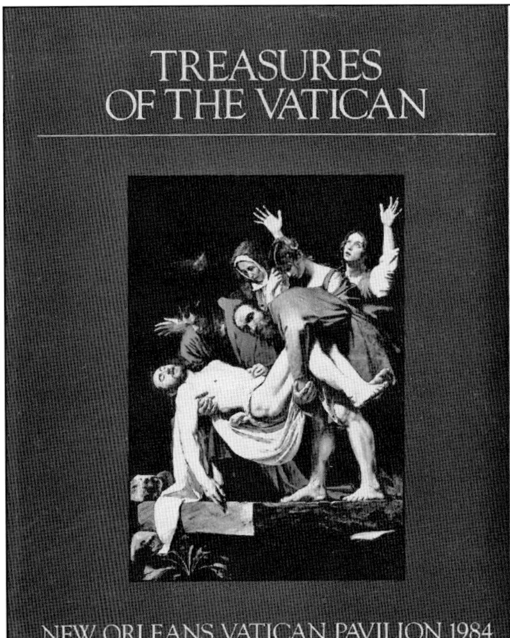

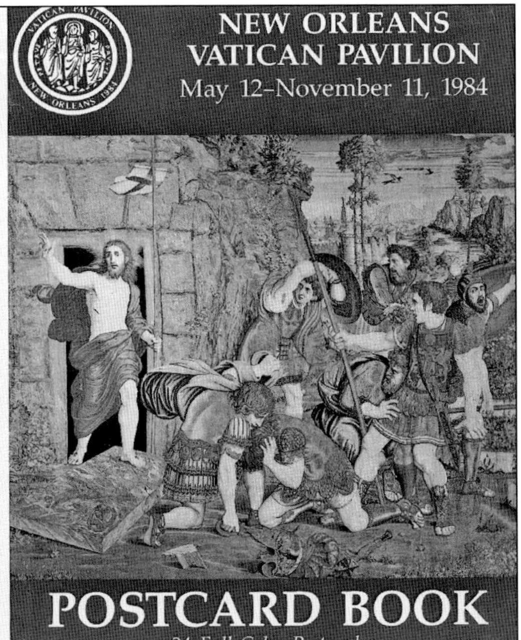

Few of the exhibitors at the fair offered pamphlets or brochures about their exhibits, but the Vatican Pavilion sold two booklets with views of the artwork on exhibit. The book on the left is still in demand today due to the close-up pictures of many of the pieces, which makes it an invaluable tool for art students and historians.

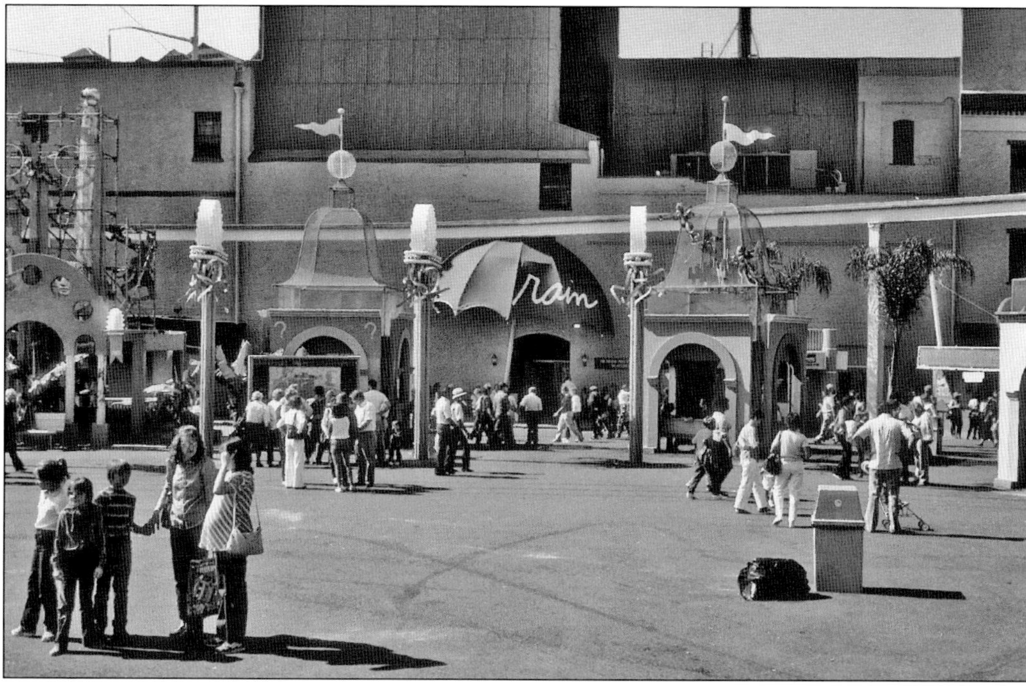

Located in the rear of the mall facing Centennial Plaza, the City of New Orleans/Historic New Orleans Collection Pavilion offered an exhibit aptly titled *Rain*. Photographs, models, and sculptures were used to show the effects rain has on New Orleans, from affecting the moods of its citizens to creating special challenges due to the city's low terrain.

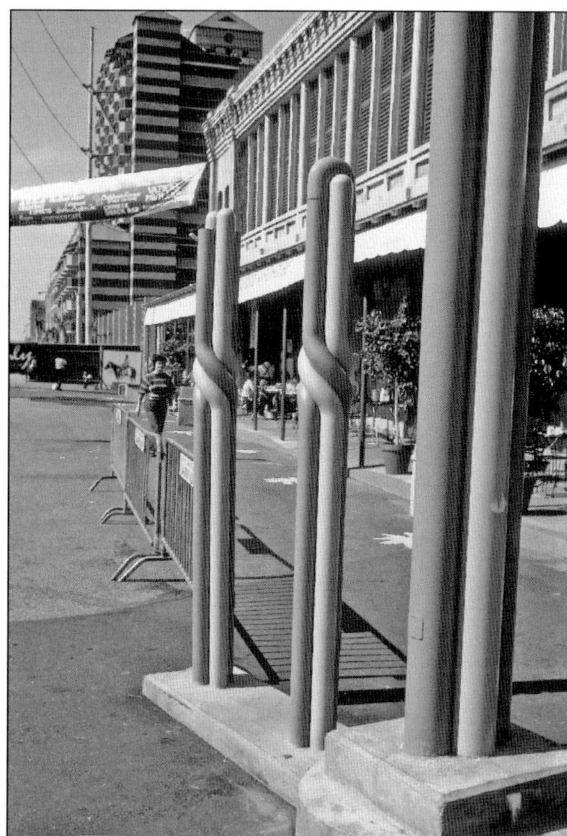

In an effort to break up the harsh industrial lines of the area, fair designers tried little touches like this set of colorful pipes twisted into unusual shapes.

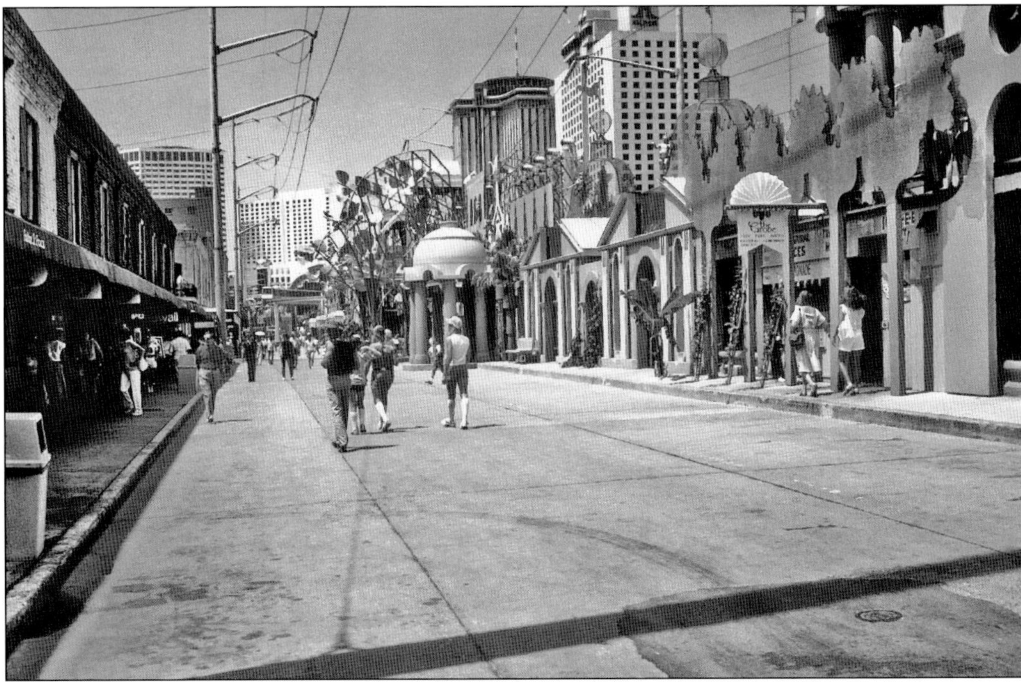

A more ambitious attempt to disguise the industrial nature of the area was the critically acclaimed Wonderwall, a fanciful collection of shapes and colors that is more fully described in chapter eight.

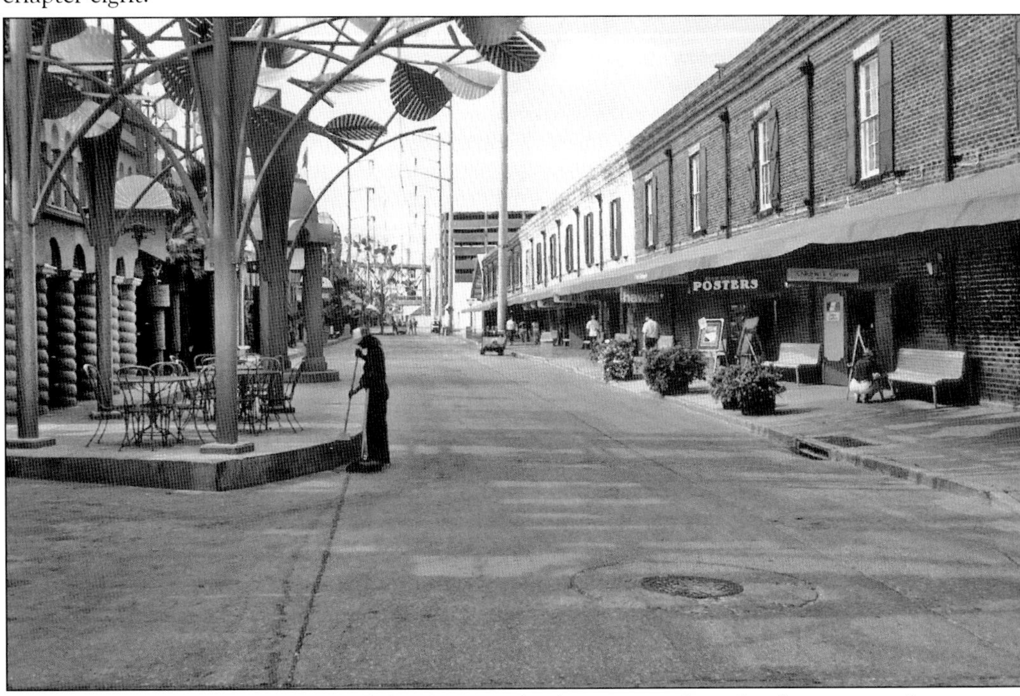

The Wonderwall and the rest of the fair may be long gone, but the former warehouses and factories are now much in demand. The tourist-oriented concessions in the mall have given way to expensive loft apartments, upscale restaurants, and chic shops.

Four

FESTIVAL PARK

Just beyond Fulton Street is Festival Park. True to its name, it is a festive area that celebrates the good life! In Cajun, that means, "Laissez les bon temps rouler!"

—1984 Annual Report

Festival Park was another section of the site that required planners to find new uses for old buildings. Unlike the Fulton Street Mall section, some of the buildings in this area could be removed, because they were not protected by any preservation rules or they were too deteriorated to save. This allowed for the clearing of a large central area that could be filled with new structures for the fair, making it look more like an exposition and less like just another part of the city.

Overlooking the rest of the area was the massive Federal Fibre Mills building, a five-story brick-and-timber factory that was used during the fair to host the Louisiana Folklife Festival, an area set aside to demonstrate the crafts, art, and music of the host state. Guests could watch workers building small boats intended for use in the bayous and swamps, learn about alligator hunting, or see a variety of musical acts perform. The building also provided a different type of entertainment: a German beer hall and restaurant.

Off to one side of the plaza, guests could enjoy the sights and tastes of the Italian Village, which also made use of vintage buildings. Styled after Venice's Piazza San Filippo di Giacomo, the village housed most of the area's outdoor entertainment, including jugglers, puppet shows, singers, and magicians.

A decidedly different type of entertainment could be found in the Jazz and Gospel Tent. Two different stages inside hosted up to 12 acts daily from across the South as well as from Europe, Africa, and the Caribbean. Guests were invited to join in, and the tent was often filled with singing, clapping, and dancing crowds for much of the day and night. Just to round things out, Festival Park also included several cutting-edge amusement rides to keep things interesting.

It was initially planned to keep the Italian Village open after the fair, but declining business finally led to its closure as well. The open part of the plaza is now a city park, and the Federal Fibre Mills Building has been converted to condominiums.

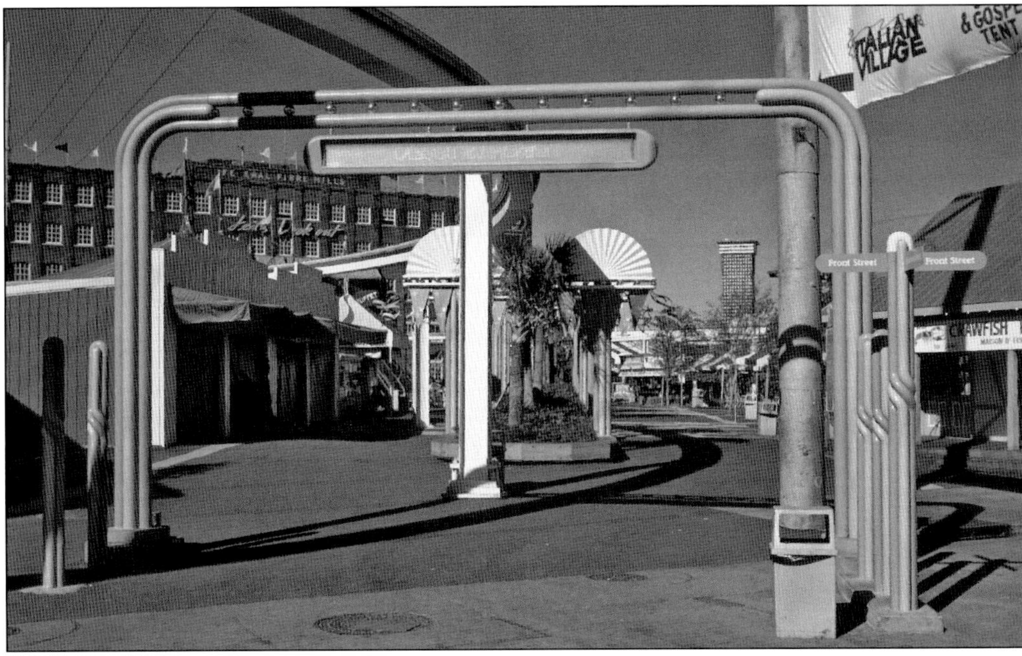

Like the Fulton Mall area, Festival Park was a strange mixture of styles. Visitors had to reconcile varying design elements, such as this futuristic walkway, with older structures such as the historic Federal Fibre Mills Building in the background. In this case, there was more open area, and thus the design was somewhat more successful.

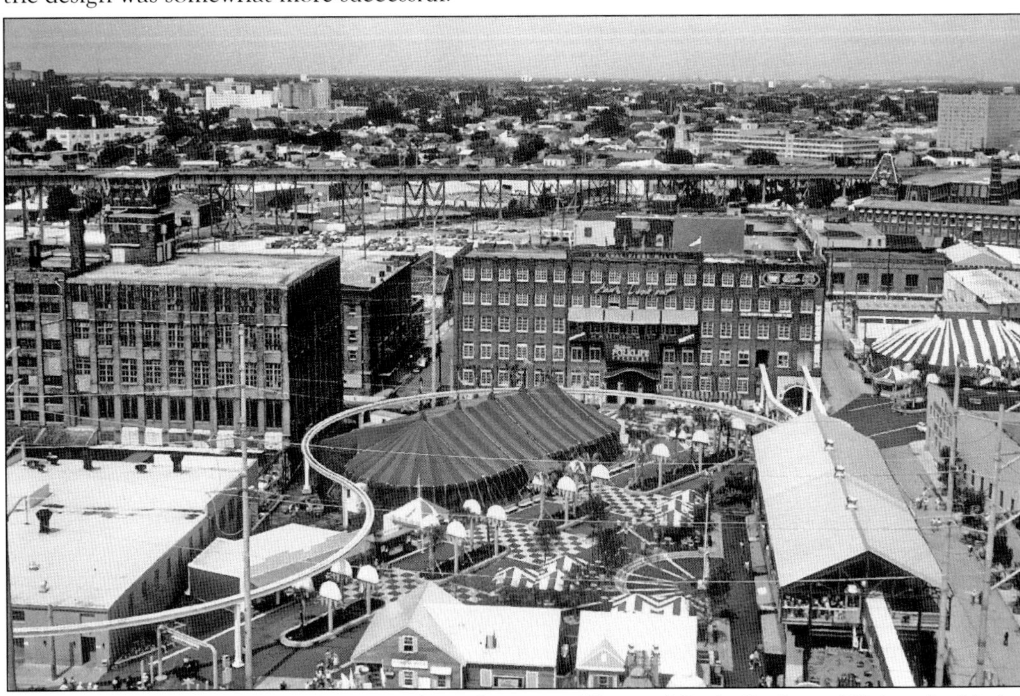

The Federal Fibre Mills Building stands at the rear of Festival Park. Built in 1904 and added to the National Register of Historic Places in 1983, the building was originally used as a rope factory. For the fair, it housed art exhibits, snack bars, shops, and a beer garden.

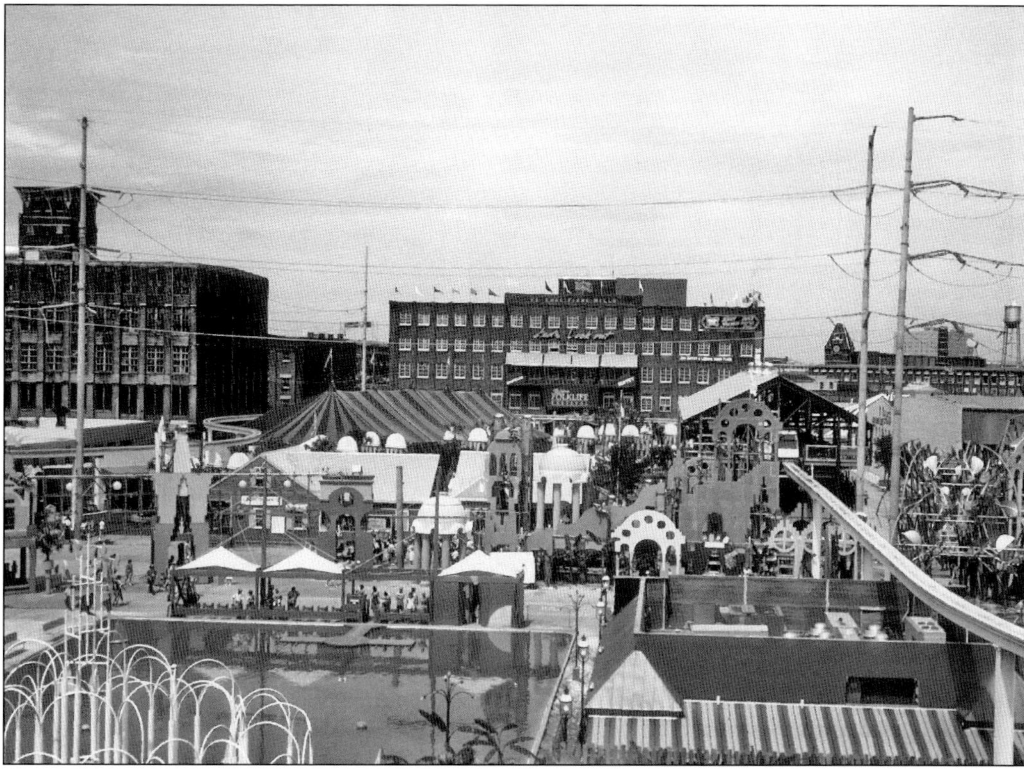

This view from the Giant Wheel across the Watergarden towards the Federal Fibre Mills Building shows how varied the architecture was in Festival Park. While small compared to efforts such as the 1964–1965 New York World's Fair or Montreal's Expo 67, there was still a considerable amount of ground to cross, leading to one thing common to all world's fair—sore feet.

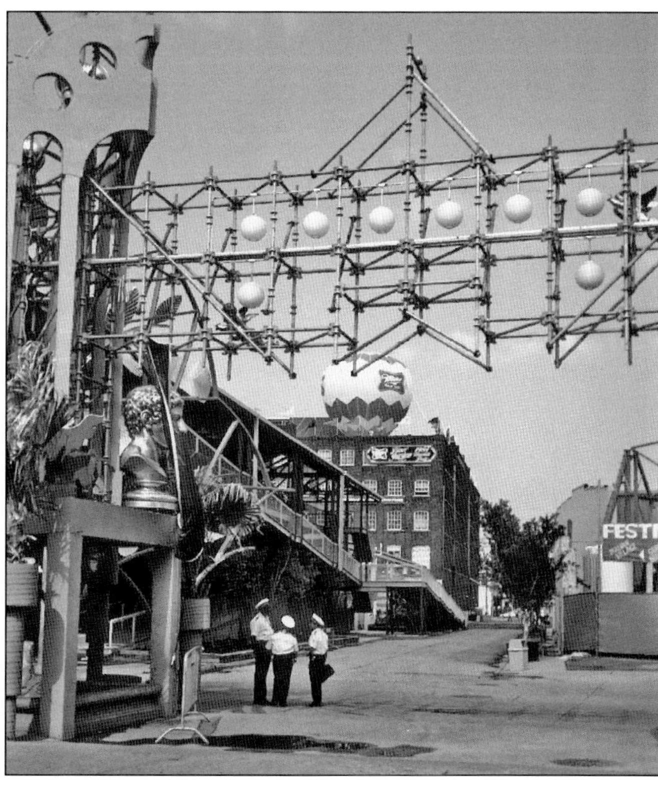

Although Germany was not one of the nations participating in the fair, guests could feel like they were there with a visit to the Miller Beer Garden. A salute to the famous Oktoberfest celebration, the offerings included Bavarian food and music as well as plenty of beer.

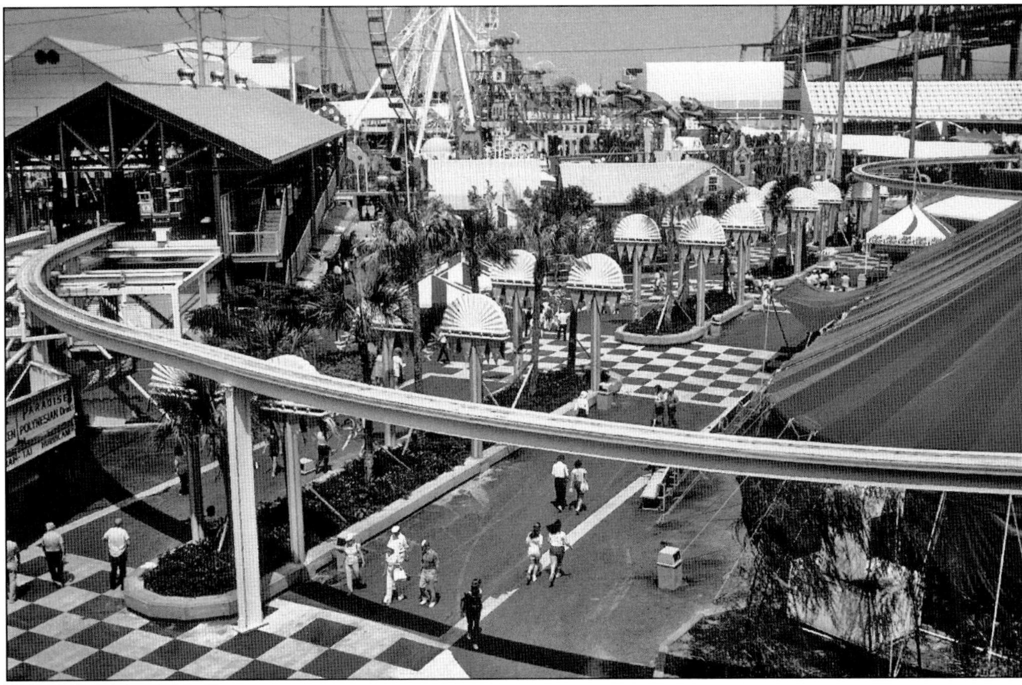

Festival Park was another one of the stops on the monorail system. The checkerboard pattern on the ground between the monorail station and the Jazz and Gospel Tent brightened up the area and was a definite improvement over the blank concrete expanses found in much of the fair.

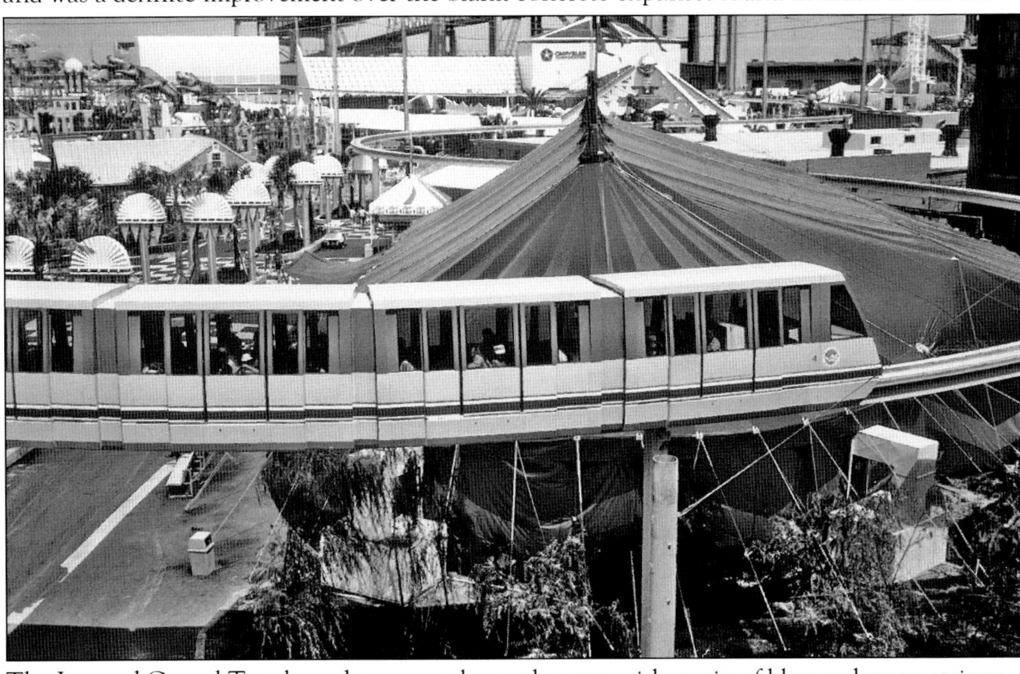

The Jazz and Gospel Tent brought more color to the area with a mix of blue and green stripes. A tent was the ideal design for the venue, as it captured the Southern tradition of revival meetings and traveling ministries. A wide variety of musical styles provided by groups from across the state made this a popular destination with local residents as well as tourists.

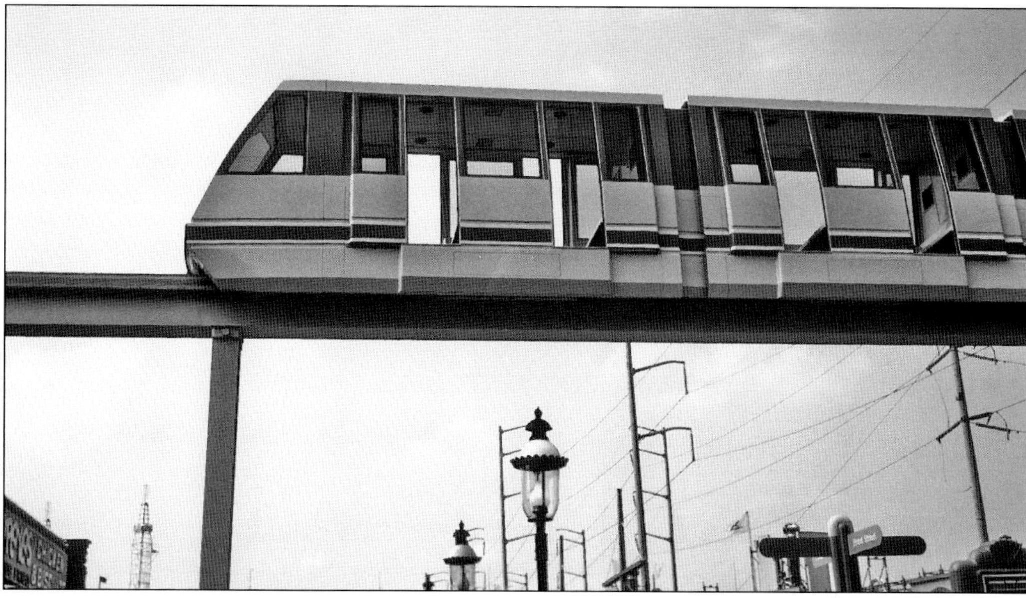

The air-conditioned monorail was a popular way to see the fair. Each train consisted of 10 cars traveling on a 1.6-mile-long beam that crossed much of the grounds. Computer controlled, the monorail operated without human drivers. The doors would not normally have been open outside the stations, as seen in this picture of a train undergoing maintenance.

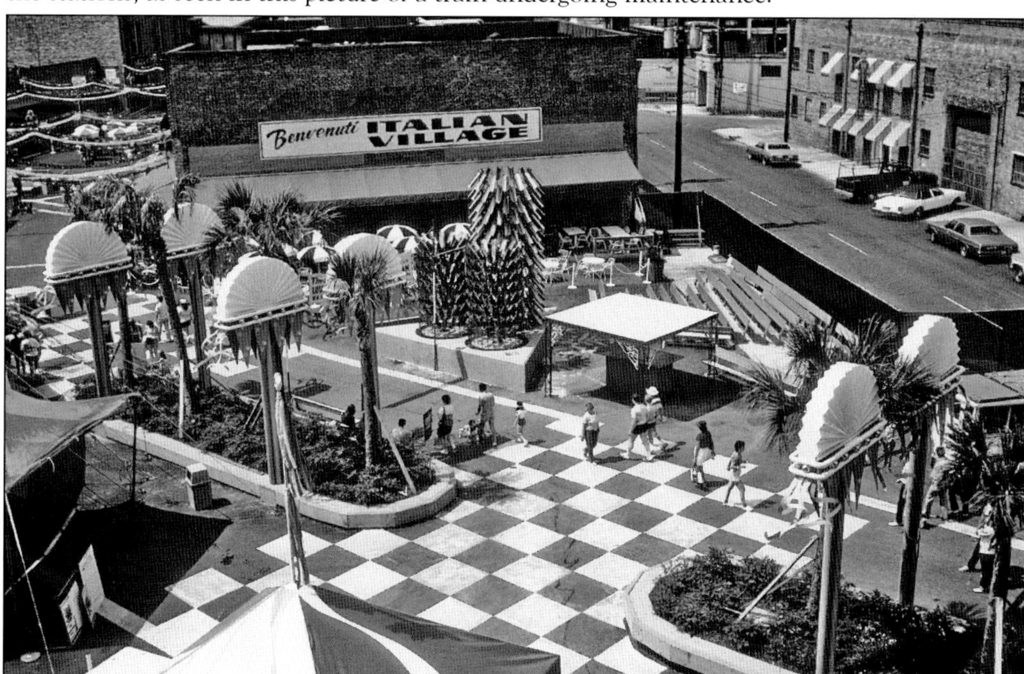

One of the major attractions in Festival Park was the 60,000-square-foot Italian Village. The village was a major undertaking, complete with shops, restaurants, a tavern, a museum, and even an art studio where resident artists entertained the crowds as they worked. After the fair closed, many of the displays and the artwork were moved to the new American Italian Renaissance Foundation Museum, where they can be seen today.

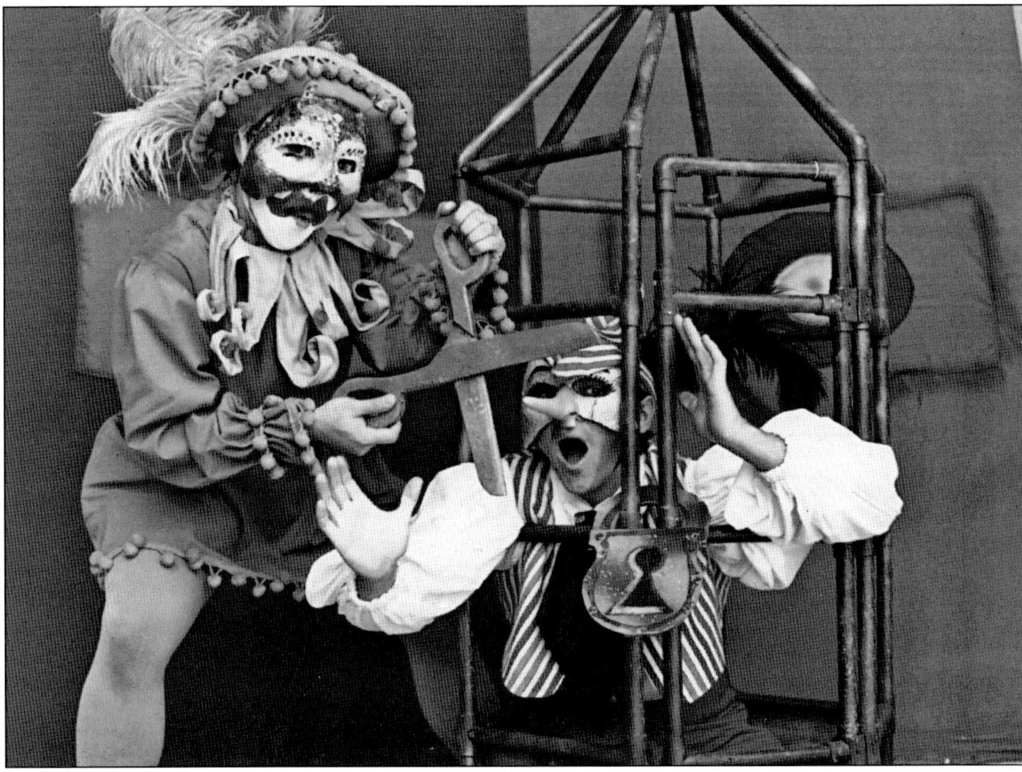

The Italian Village also hosted a large number of street performers in the open area, with dancers, mimes, jugglers, and magicians entertaining passersby. Children were treated to a daily birthday party for Pinocchio as well as puppet shows and a chance to appear onstage in colorful costumes. (Courtesy of Edward R. Cox.)

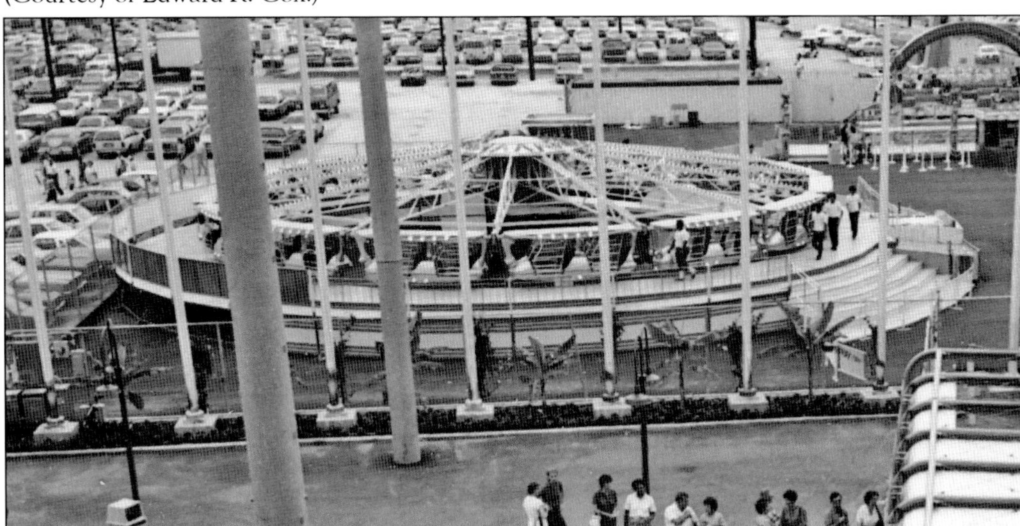

What looks like a benign and friendly merry-go-round was actually a much more challenging ride. Sky Lab consisted of 15 suspended gondolas that would swing out as the ride picked up speed; then the whole assembly would rise and tilt, reaching a height of 85 feet as centrifugal force pinned riders to their seats.

Five

Bayou Plaza

Cross the Wonderwall again, and you are in Bayou Plaza, near the Bridge Gate, a world of water.

—1984 Annual Report

After dealing with the issues caused by having to preserve the historical buildings in the Fulton Street Mall and Festival Park sections, the designers must have enjoyed themselves with Bayou Plaza. The plot could be cleared and new buildings constructed just for the fair, making this section one of the fair's more fanciful ones.

Bayou Plaza was one of the main entrances to the fair and primarily was used by those arriving by car or bus. The fair experience began as guests arrived at the fanciful Bridge Gate, which was guarded by an angry-looking sea god and a trio of alligators. Some parts of Bayou Plaza would have seemed right at home at other fairs. The Aquacade, for example, was very similar to the one built for the 1939–1940 New York World's Fair, complete with synchronized swimmers and high divers. A vintage locomotive that had previously been seen at Expo 74 was on display, close to a Ferris wheel, which was introduced at an earlier world's fair and had been part of many others over the years. Chrysler's display of its newest cars and trucks continued a long tradition of automotive displays at American fairs.

The area also contained displays and architectural touches that were unique to this fair. The very name of the section was intended to evoke thoughts of the bayous and waterways that make up much of Louisiana, so a large lagoon was a key element in the design. A wooden walkway made to look like a wooden fishing pier flanked one side of the lagoon, with the strange shapes of the Wonderwall on the opposite shore. Special lighting on the Wonderwall and Ferris wheel provided a colorful spectacle each evening, continuing the fun as weary guests made their way back towards the parking lots.

After the fair ended, all of the structures in Bayou Plaza were removed, initially leaving the area empty, because it was earmarked for possible extension of the convention center. That facility has now been expanded, completely eliminating any signs of this part of the fair.

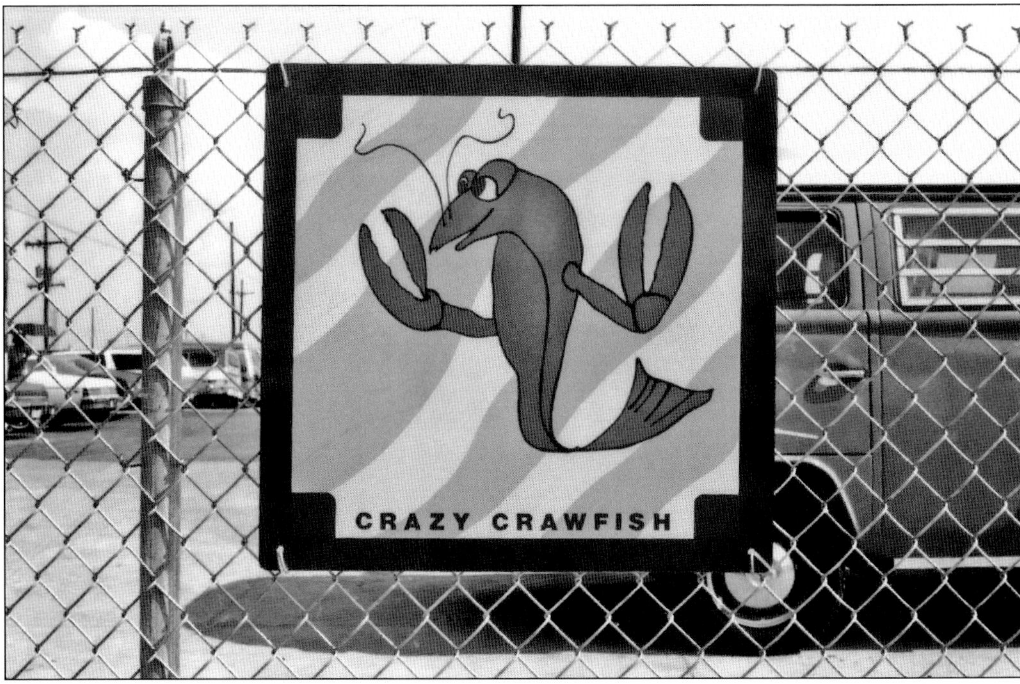

Bayou Plaza was the main entry point to the fair for guests arriving by car or bus. The fun started with the signs designating the sections of the parking lot, for the location signs sported cartoon characters based on local themes.

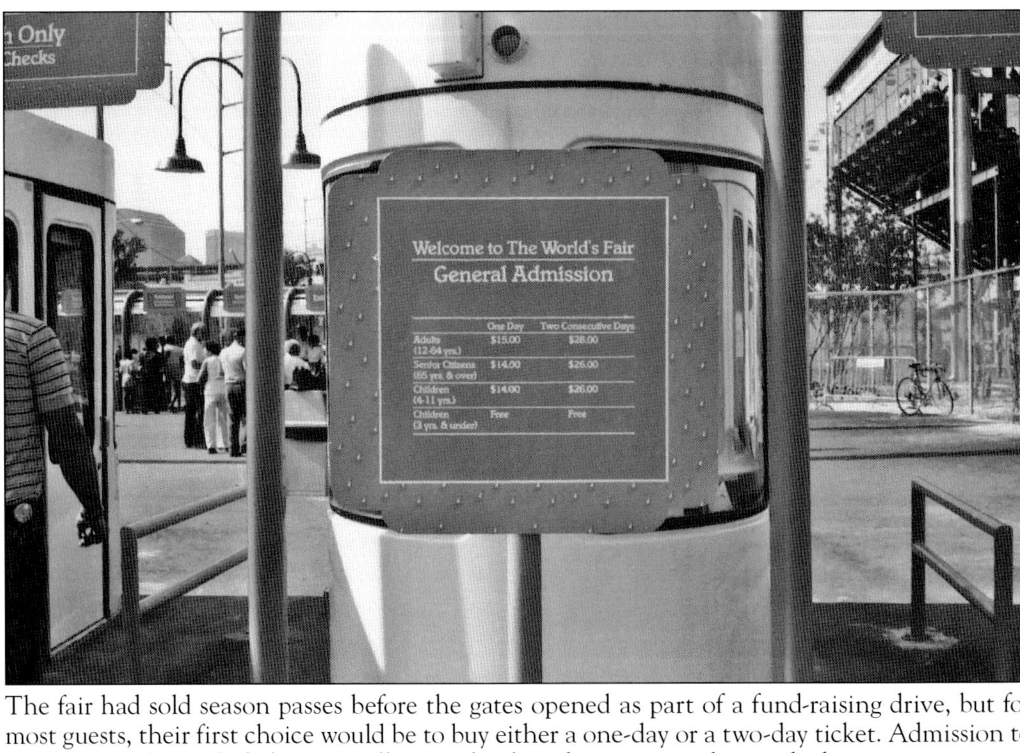

The fair had sold season passes before the gates opened as part of a fund-raising drive, but for most guests, their first choice would be to buy either a one-day or a two-day ticket. Admission to the fair was also included in specially priced tickets for events at the amphitheater.

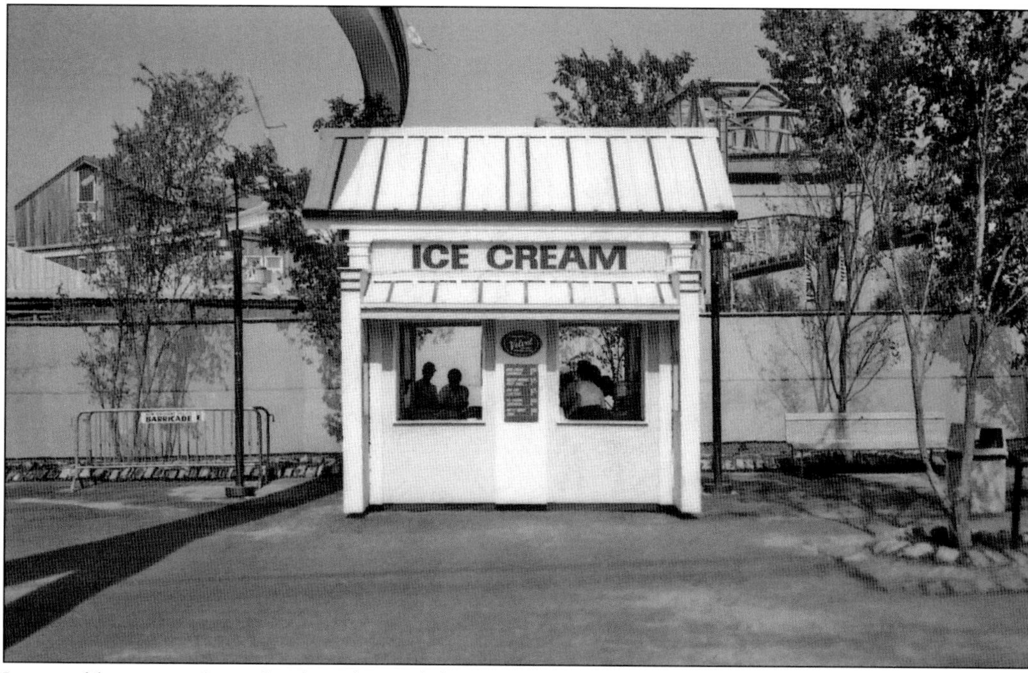

Located between the ticket booths and the turnstiles, this modest-looking stand may have been empty in this early-morning view, but when the day started heating up, it became very popular with crowds already hot from their walk in from the parking lot.

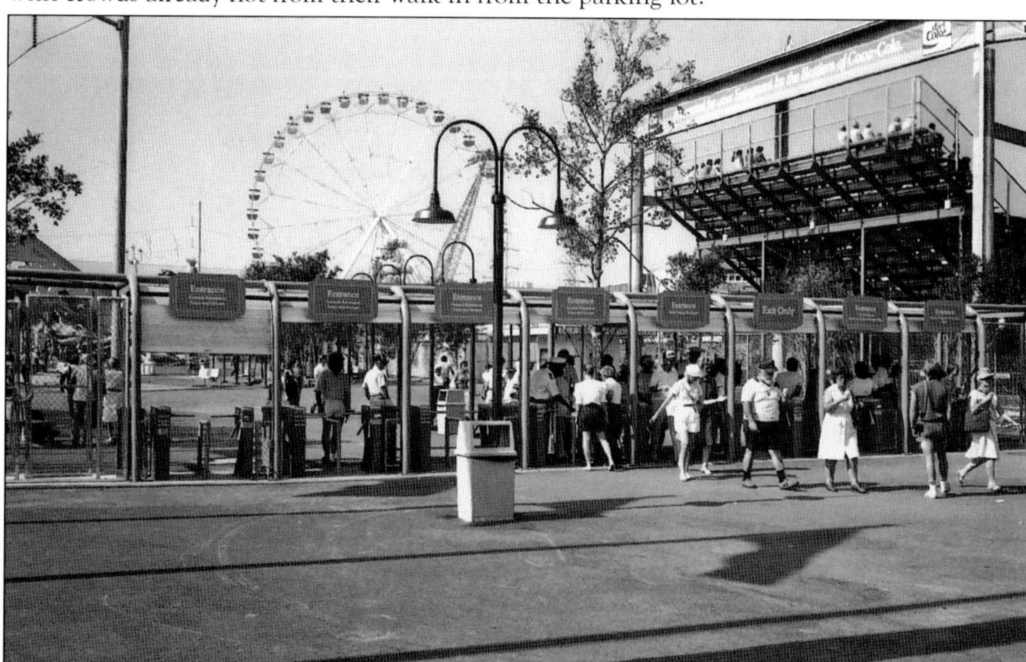

As guests arrived at the turnstiles, they got their first glimpse of the fair itself. In only a few more moments they could be heading towards the Giant Wheel looming in the background, splashing about in the Watergarden on the way, or stopping to enjoy a show in the Aquacade just off to the right.

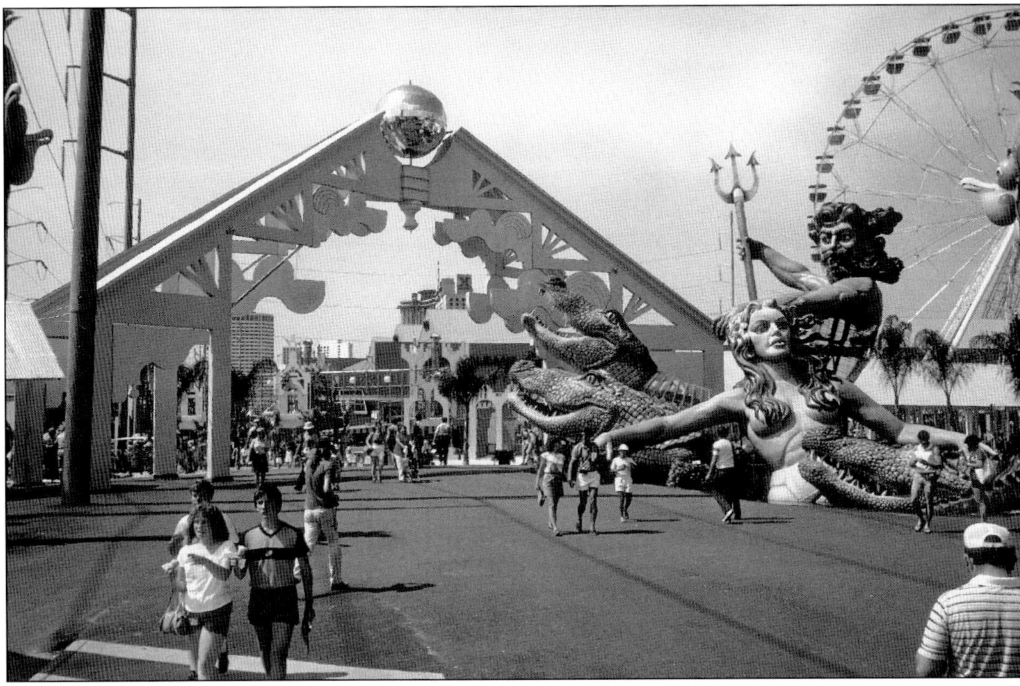

After passing through the turnstiles, guests arrived at an imposing archway named Bridge Gate. The gate was guarded by oversized figures including a voluptuous mermaid, pelicans, alligators, and an angry-looking King Neptune. Like the figures at the City Gate, the eye-catching mermaid probably helped sell more film.

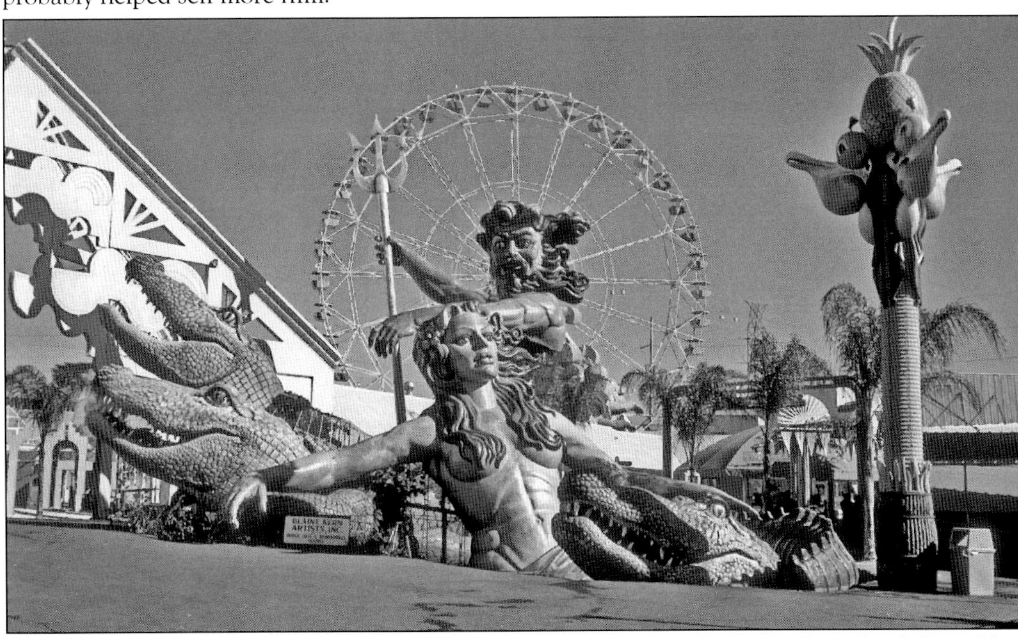

Neptune does not look too pleased to see the visitors entering the fair. Perhaps he did not like the way many of them were ogling the barely clad mermaid, or maybe he was warning them about the alligators that surrounded her. The giant figures were built by Blaine Kern, a New Orleans artist well known for his float designs for Mardi Gras.

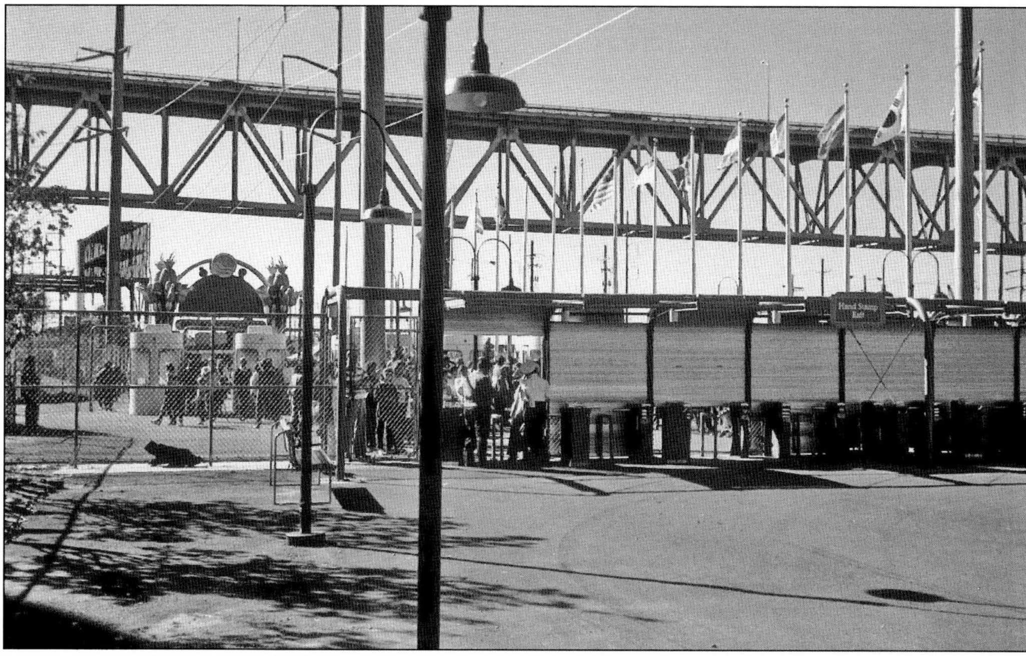

This view looking towards the Bridge Gate turnstiles shows how this entrance got its name. The Greater New Orleans Bridge stretching overhead was in the midst of a major expansion project during the fair, and it remains a major arterial link in the city's highway network. The twin spans were renamed in 1989 as the Crescent City Connection.

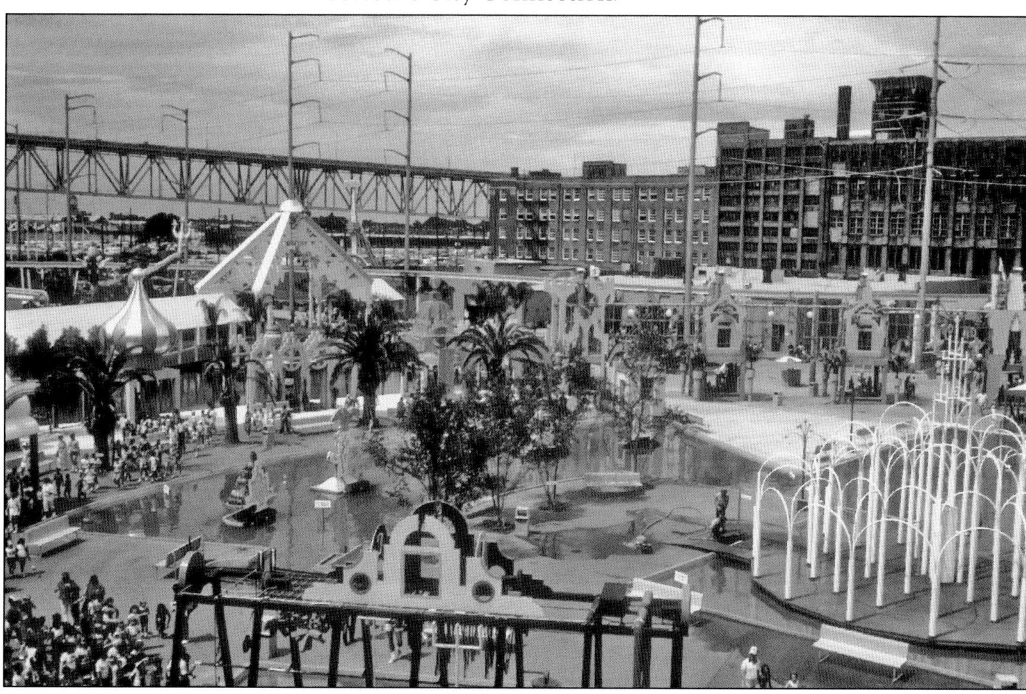

The bright colors of Bayou Plaza were a striking contrast to the walls of the former factory buildings in the background. This section of the plaza was the Watergarden, a collection of 14 fanciful fountains that built upon the fair's rivers theme.

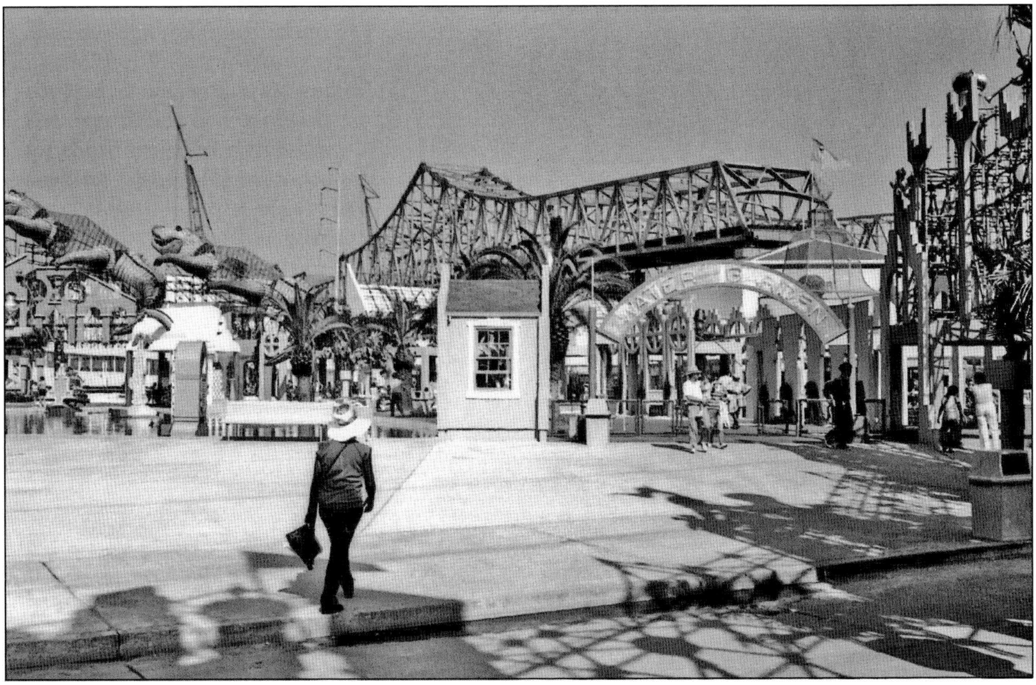

The entrance to the Watergarden was overshadowed by the span being added to the Greater New Orleans Bridge. The new bridge was supposed to have been ready for the fair, but due to lengthy construction delays, it did not open until 1988, which provided additional room for the fair.

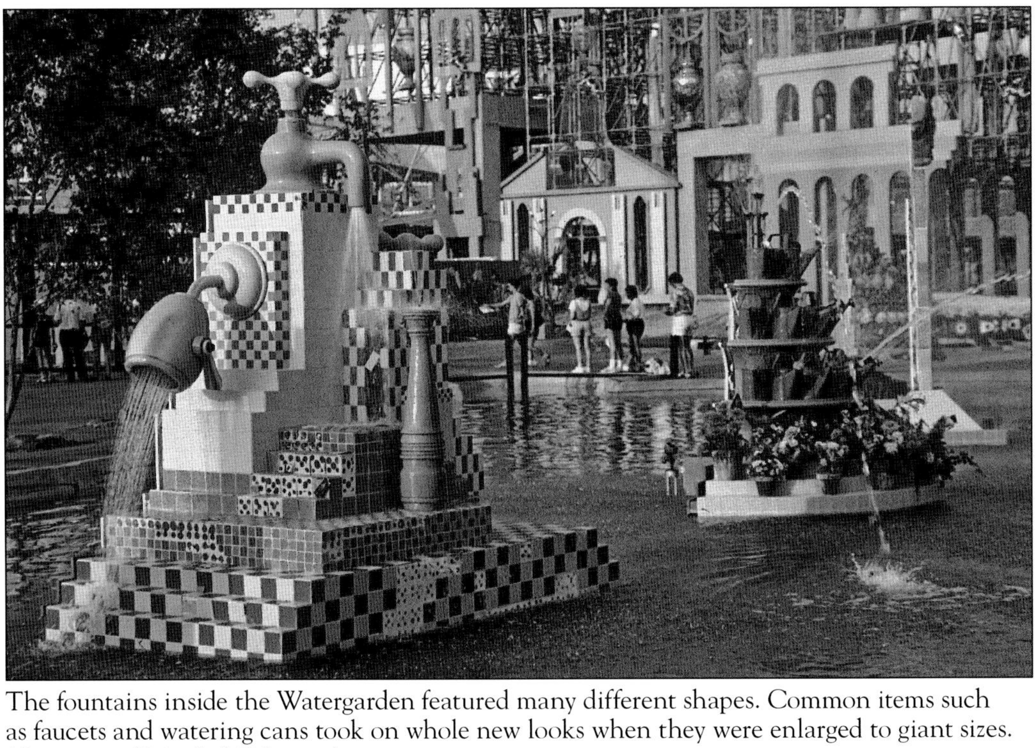

The fountains inside the Watergarden featured many different shapes. Common items such as faucets and watering cans took on whole new looks when they were enlarged to giant sizes. (Courtesy of Mitchel Osborne.)

The most popular attraction inside the Watergarden was the Kid Wash, which recreated the experience of a car wash for those brave enough for a soaking. Originally intended just for children, the Kid Wash also became popular with overheated adults looking for some fun. (Courtesy of Mitchel Osborne.)

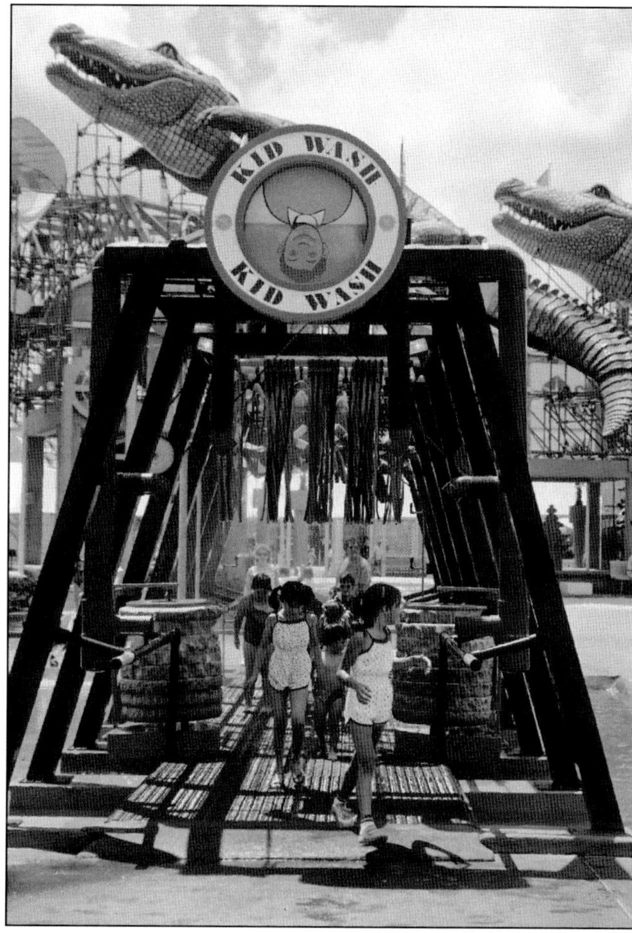

The Captain's Bridge allowed guests to pilot coin-operated boats by radio control in a 7,000-square-foot pond near the Great Hall. Afterwards they could enjoy a meal at this Popeye's Chicken and Biscuits restaurant, one of several at the fair.

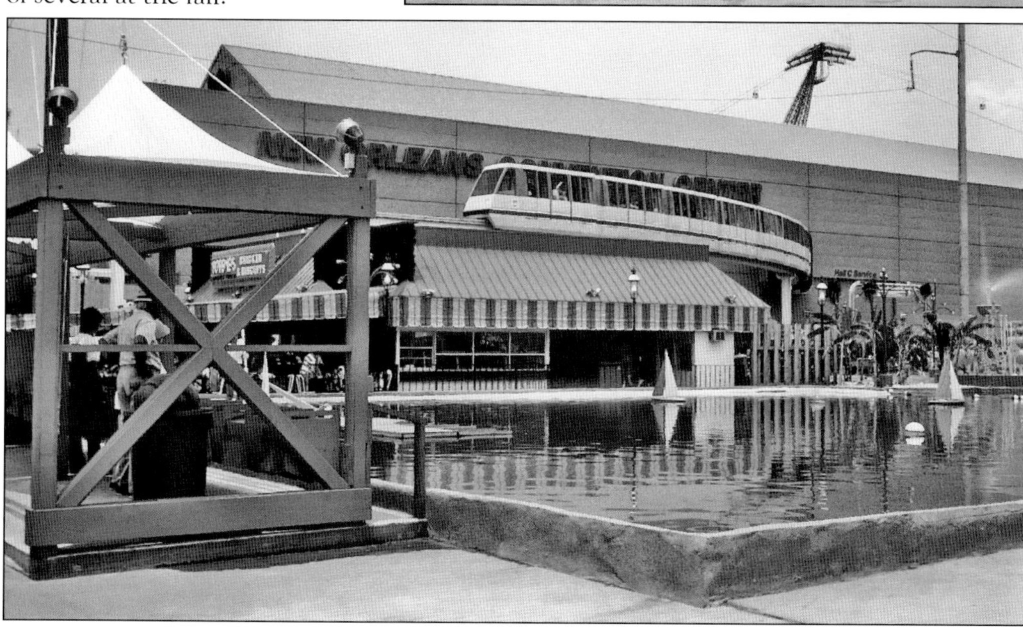

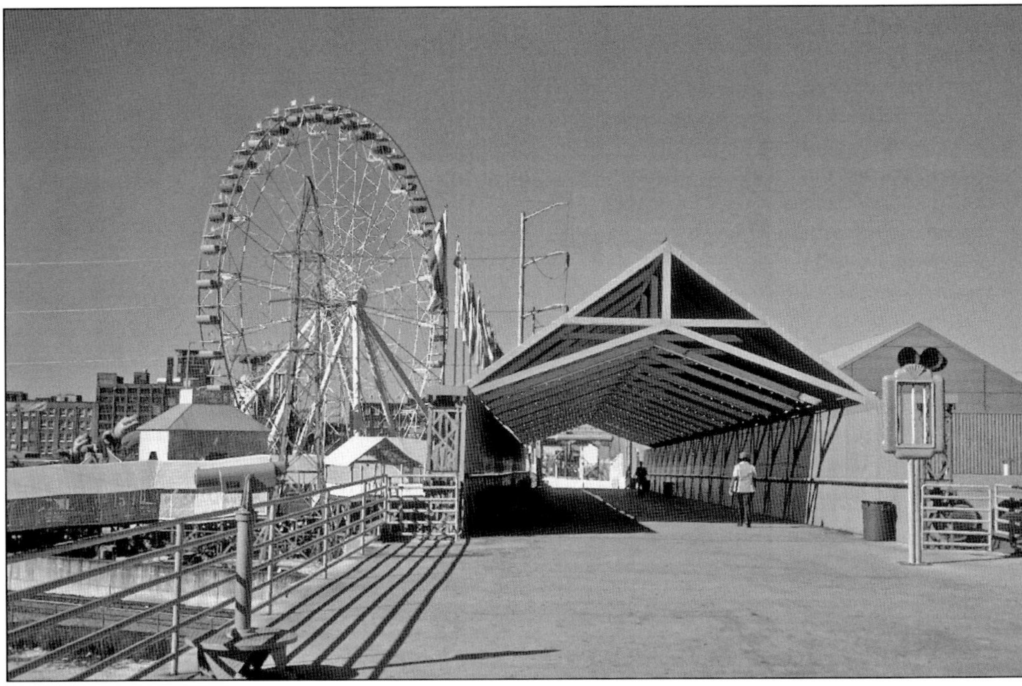

Bayou Plaza was also home to the Giant Wheel, the tallest Ferris wheel in America at the time, reaching 178 feet high. A total of 320 people could be seated in its 40 gondolas, which offered some of the best views of the fair. The Giant Wheel was completed just days before the fair opened.

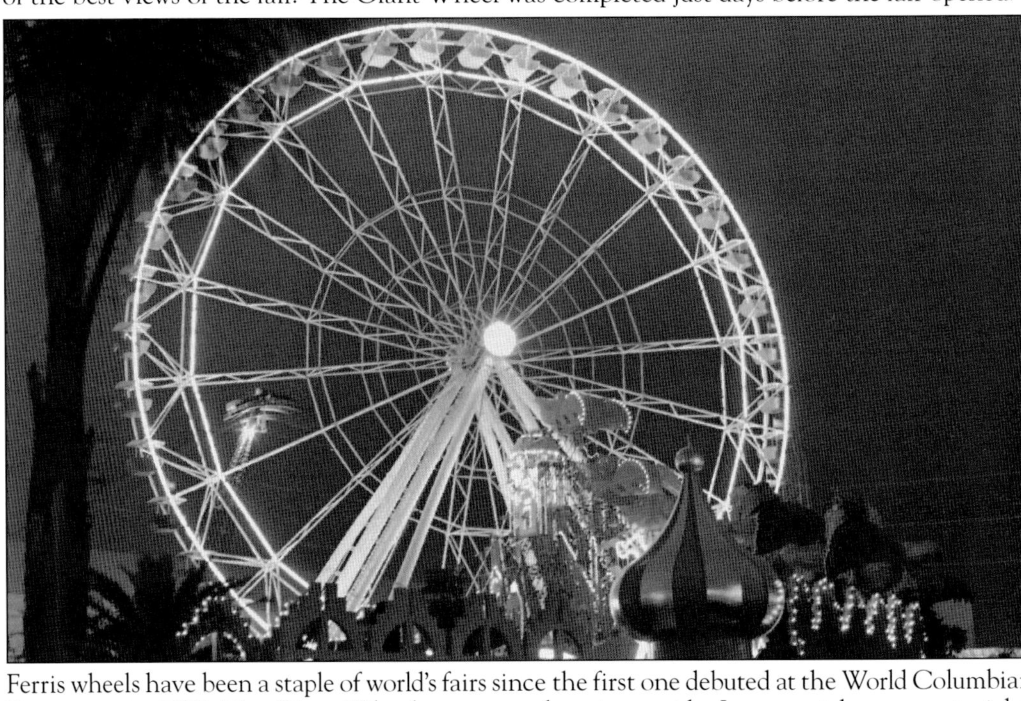

Ferris wheels have been a staple of world's fairs since the first one debuted at the World Columbian Exposition in 1893. The Giant Wheel was more than just a ride. It was a sight to see at night, when more than 4,000 lights came to life in a variety of patterns and colors. (Courtesy of James Pilet.)

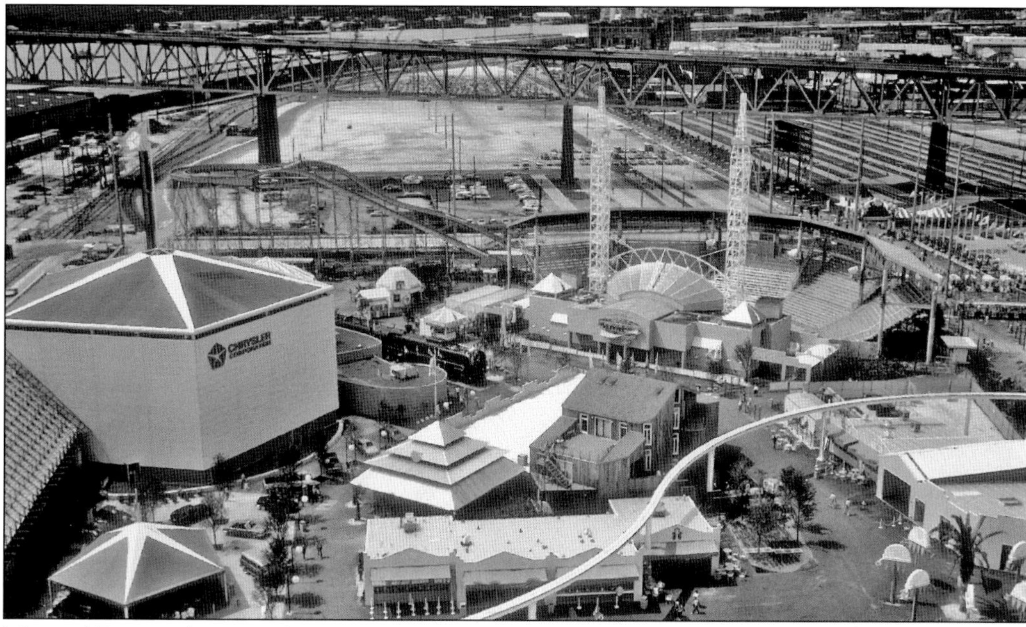

Most of Bayou Plaza can be seen in this view from the Giant Wheel. The 20,000-square-foot Chrysler Pavilion at the left was the largest commercial exhibit at the fair. The Aquacade is in the upper right, just above the Conenergy Energy Saving House, which showed some of the latest innovations in residential appliances and energy-saving designs.

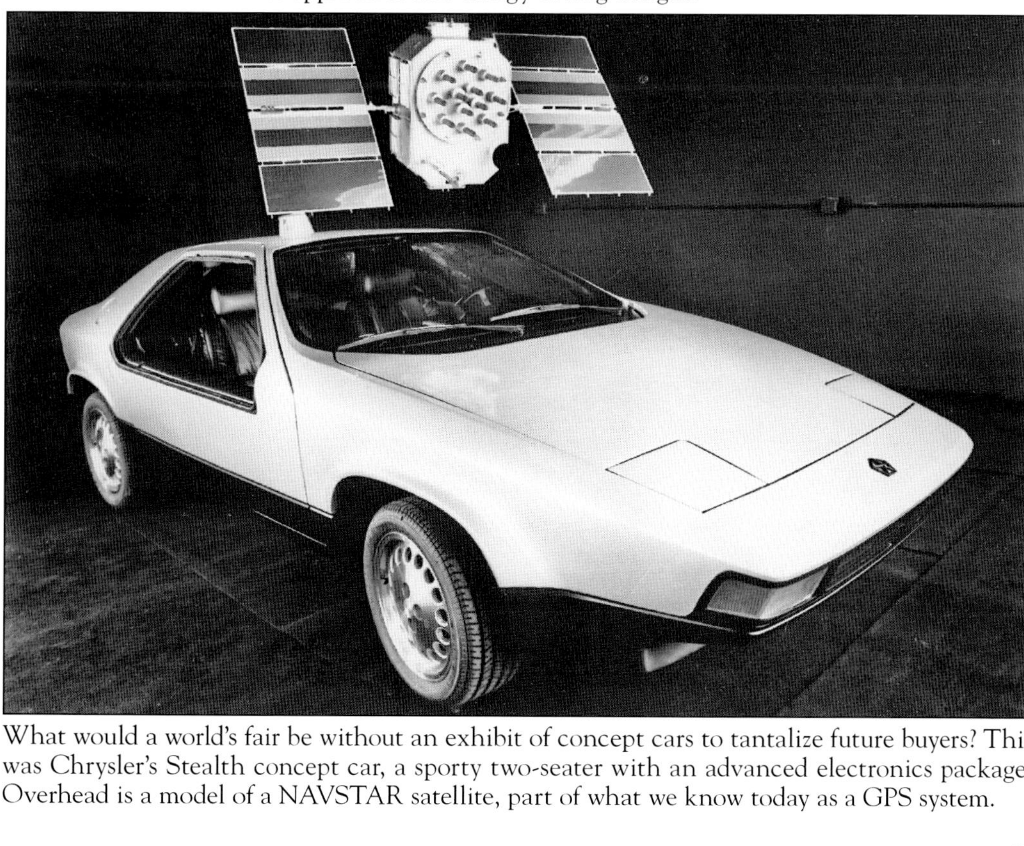

What would a world's fair be without an exhibit of concept cars to tantalize future buyers? This was Chrysler's Stealth concept car, a sporty two-seater with an advanced electronics package. Overhead is a model of a NAVSTAR satellite, part of what we know today as a GPS system.

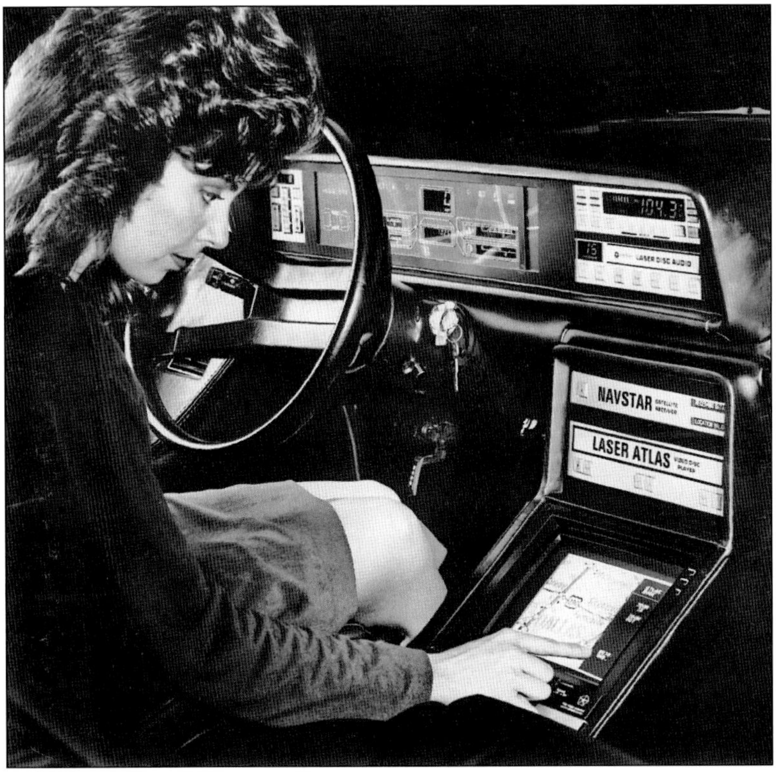

The Chrysler Laser Atlas and Satellite System used laser discs with pre-loaded maps as part of a digital dashboard. Touch-screen controls allowed drivers to locate maps as well as photographs of their intended designations and to track their progress en route.

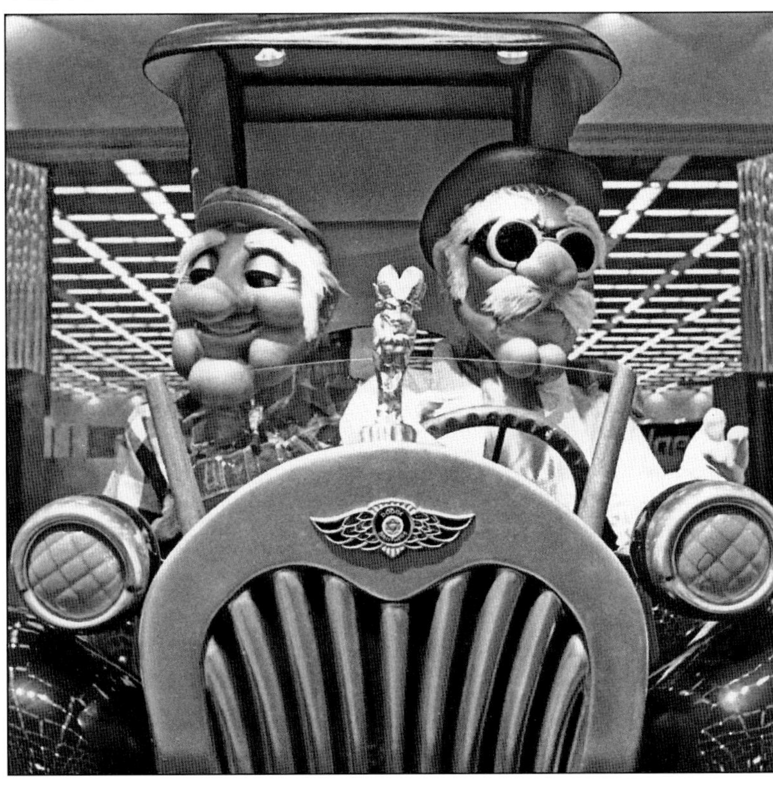

Chrysler combined some company history with a look at its new product line. Robotic characters John and Horace Dodge were based on the founders of the Dodge Brothers company. The old-time car builders introduced the audience to the wonders of the Dodge Caravan and Daytona, noting the improvements made since their own early cars.

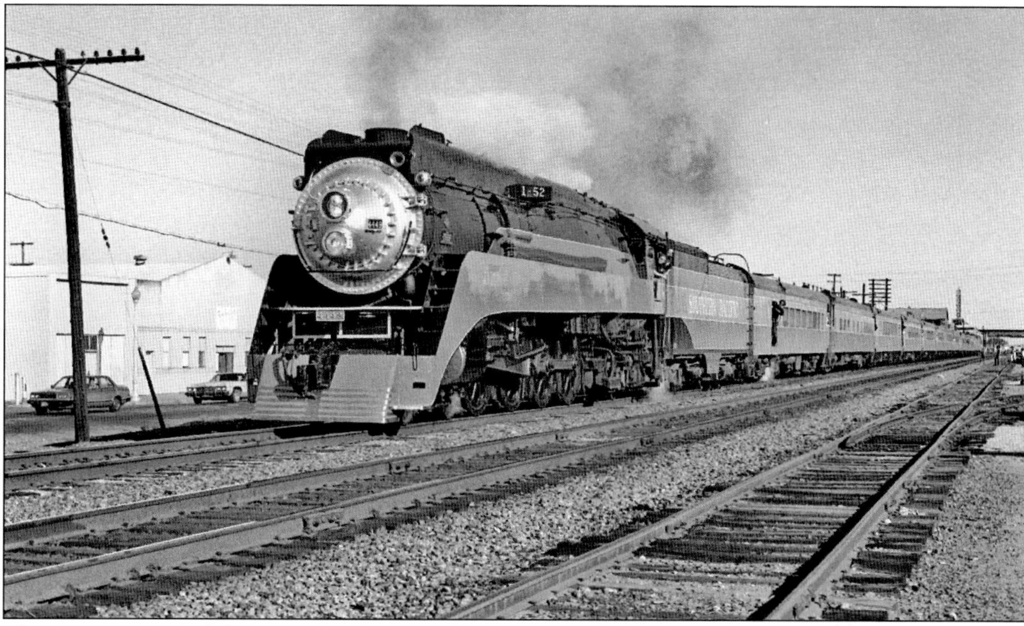

The centerpiece of the Union Pacific System Pavilion was locomotive No. 8444, the last steam locomotive purchased by the railroad. A world's fair veteran, having been displayed at Spokane's Expo 74, the engine is seen here leaving Fresno, California, on its way to the fair. The trip was so popular with steam engine fans that a book was published chronicling the journey. (Courtesy of Ivan S. Abrams.)

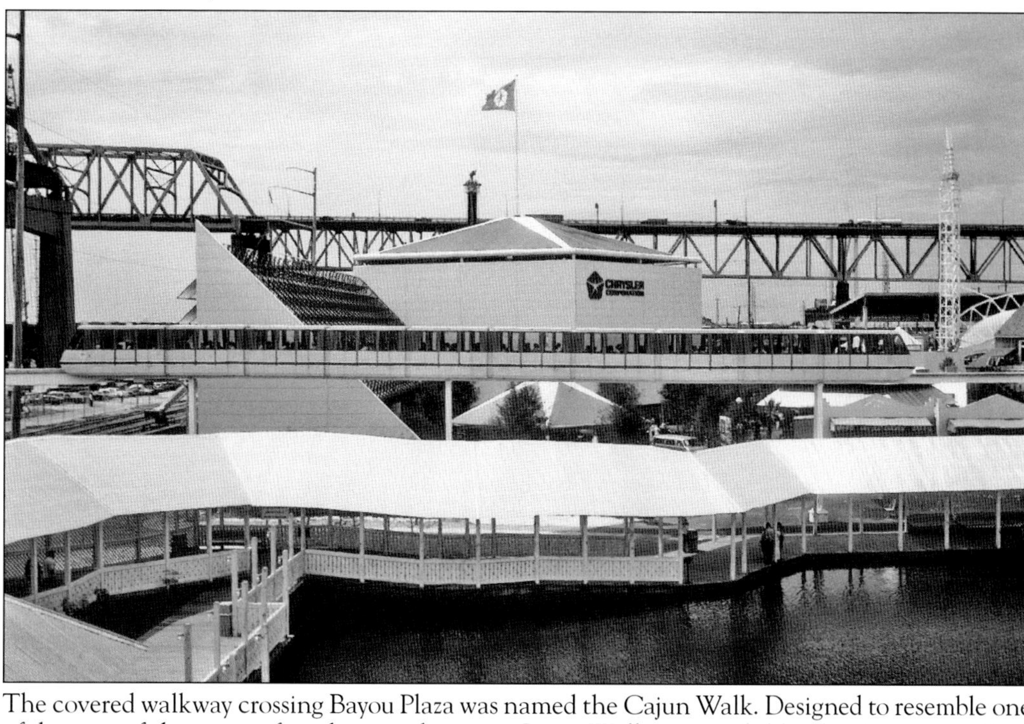

The covered walkway crossing Bayou Plaza was named the Cajun Walk. Designed to resemble one of the many fishing piers found across the state, Cajun Walk was much like the Empress Walk in Centennial Plaza, providing some shade as guests headed towards the International Riverfront.

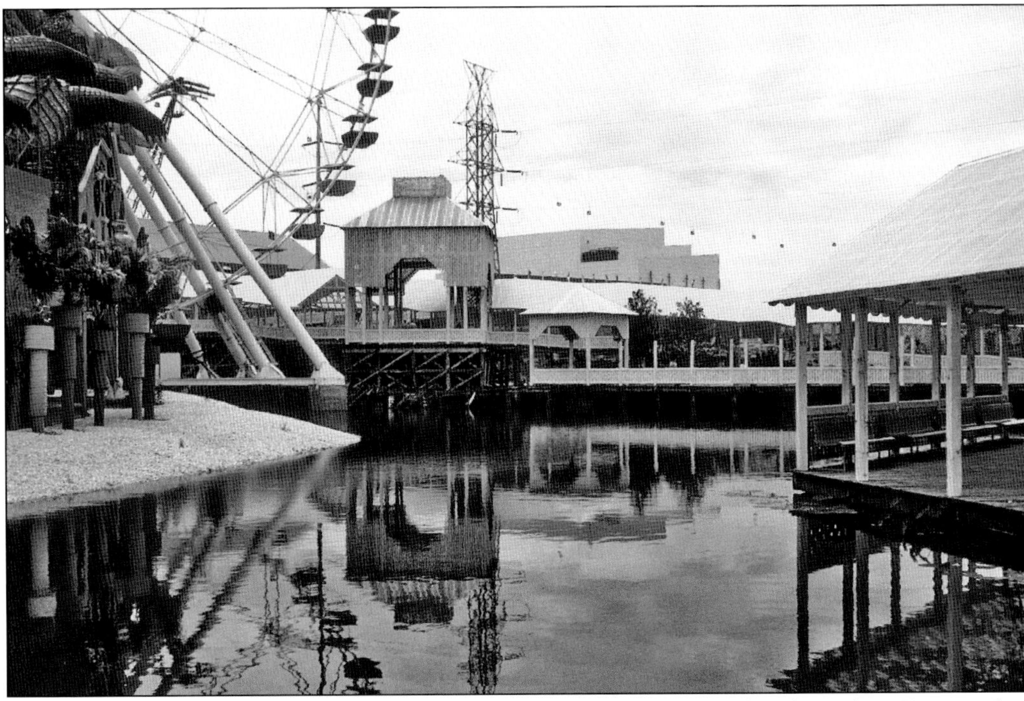

Cajun Walk skirted the edge of Bayou Lagoon, which in turn bordered on the Watergarden. The lagoon was a tranquil setting in the midst of the noise of the rest of the fair. The still water provided an extra level of interest for photographers and those just looking for a quiet place to rest for a few minutes.

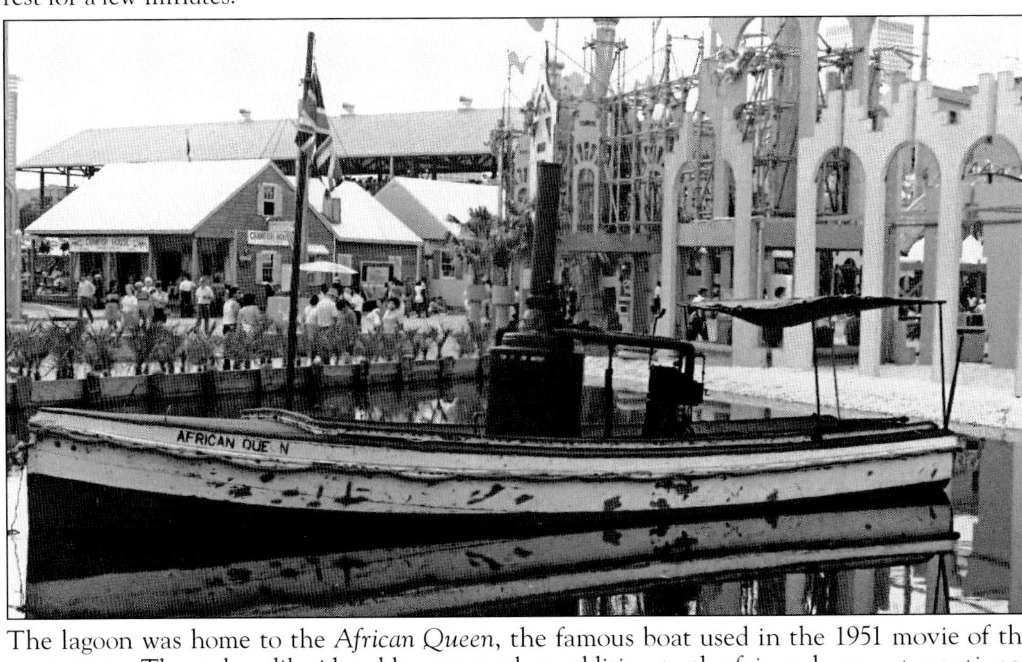

The lagoon was home to the *African Queen*, the famous boat used in the 1951 movie of the same name. The rather dilapidated boat was a late addition to the fair and was not mentioned in the official guidebook. It can be seen now on display in Key Largo, Florida. (Courtesy of James Pilet.)

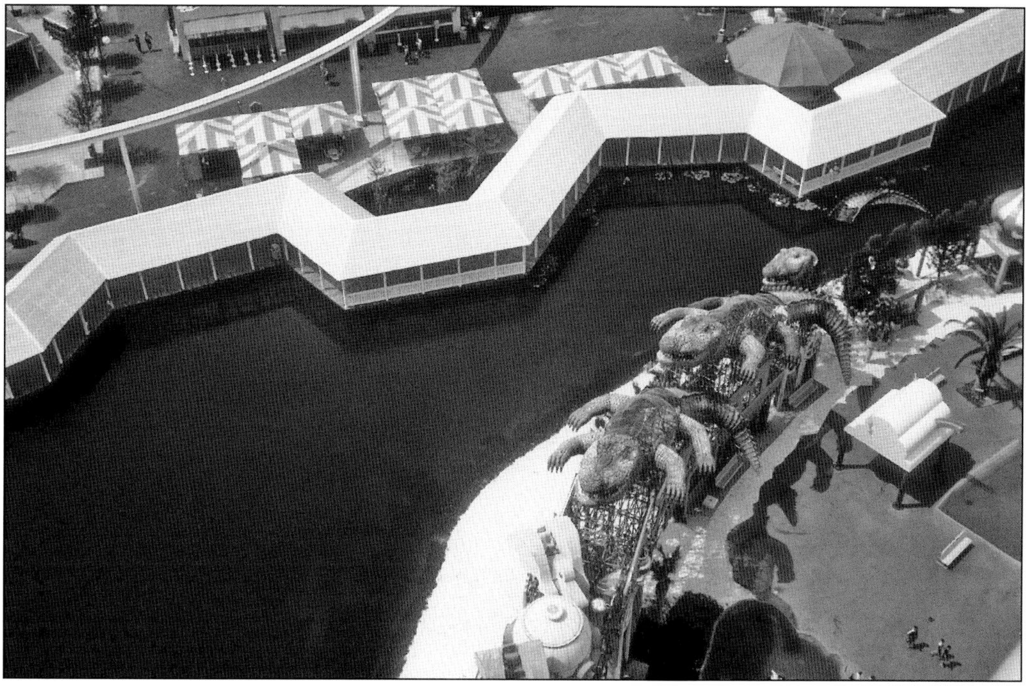

Bayou Lagoon was bordered on one side by the Cajun Walk and by part of the Wonderwall on the other. Many different Louisiana plants were located on the edges and in the lagoon itself. While it looks solid in this photograph, the roof of the Cajun Walk was built of translucent fiberglass to provide natural lighting during the day.

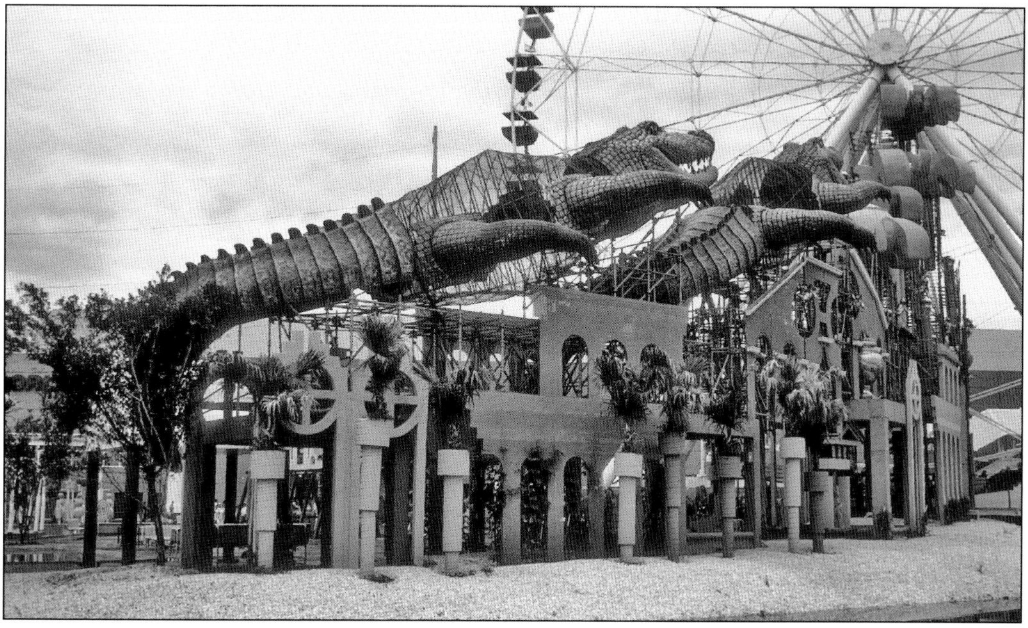

The financial woes of the fair were responsible for the half-finished alligators on top of the Wonderwall. When the money ran out during construction, they were left this way for the run of the fair. Happily, most visitors were probably unaware of the reason behind the incomplete alligators and likely attributed it all to an artistic design.

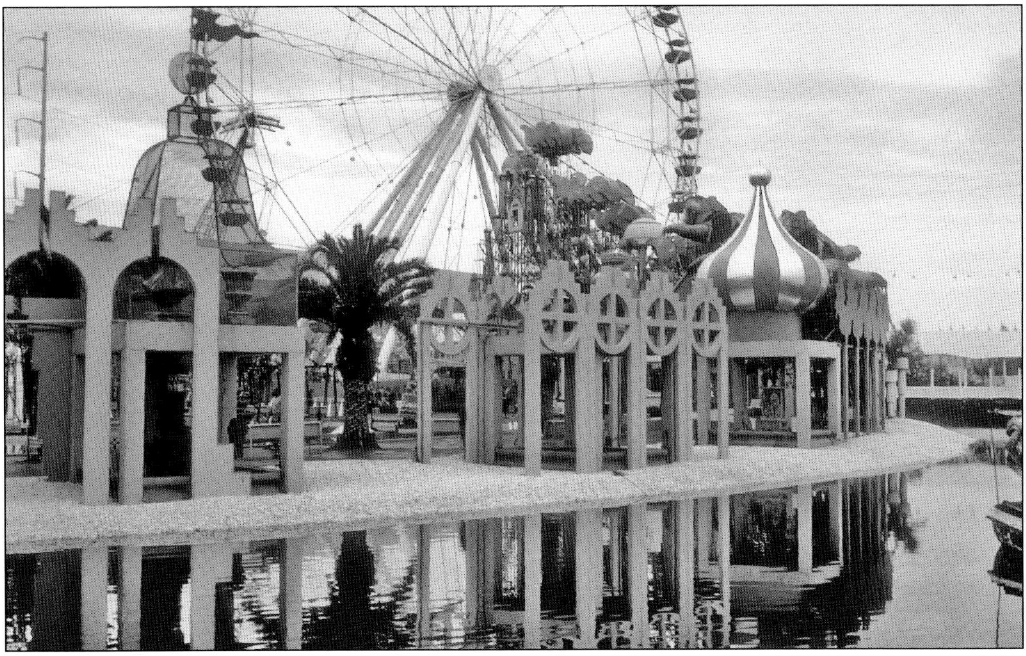

Although it may not have been readily apparent, the publicity material for this section of the Wonderwall explained that the hungry alligators were chasing after a flock of pelicans. Perhaps the pelicans were a bit too abstract in design to make that clear; they looked more like clouds than pelicans, but they were colorful and fit in well with the rest of the Wonderwall.

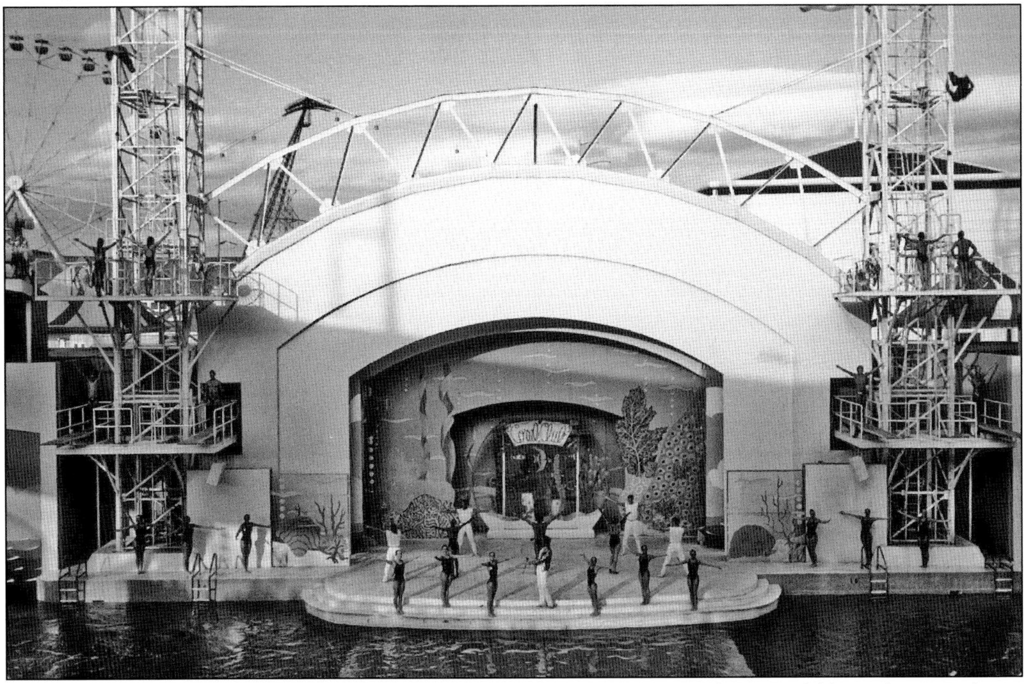

The Coca-Cola Aquacade featured *America Swims*, a water show that would have been right at home in the 1939–1940 New York World's Fair with its synchronized swimmers. Here a group of divers prepares for a high-dive act as other performers entertain onstage.

As the divers leave their platform, the Cyclo-Tower ride rises off to the right. Introduced at the fair, the ride was plagued with problems and was often closed for maintenance. Improved versions were later sold under the name Condor and were more successful.

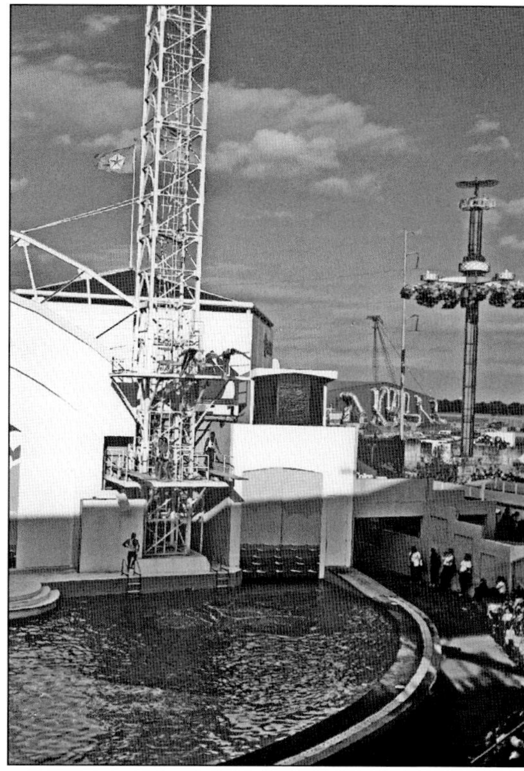

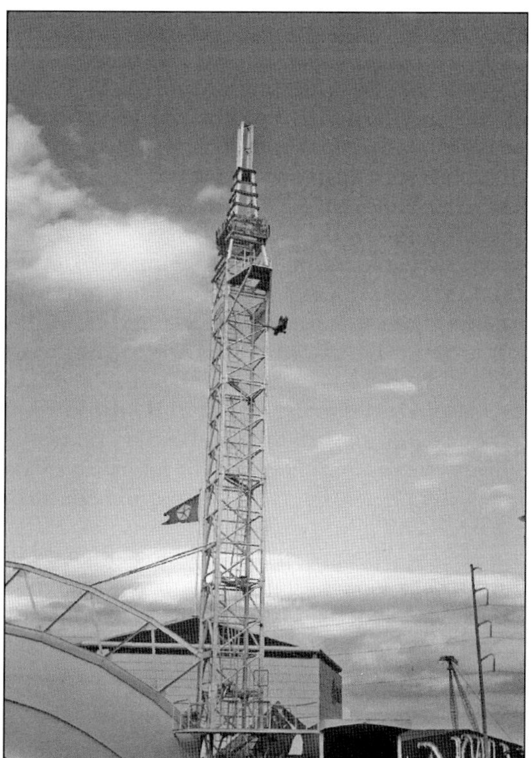

The high divers, members of the U.S. Diving Team, provided the most thrilling portion of the 35-minute show. As this view shows, their leaps were from one of the tallest structures at the fair.

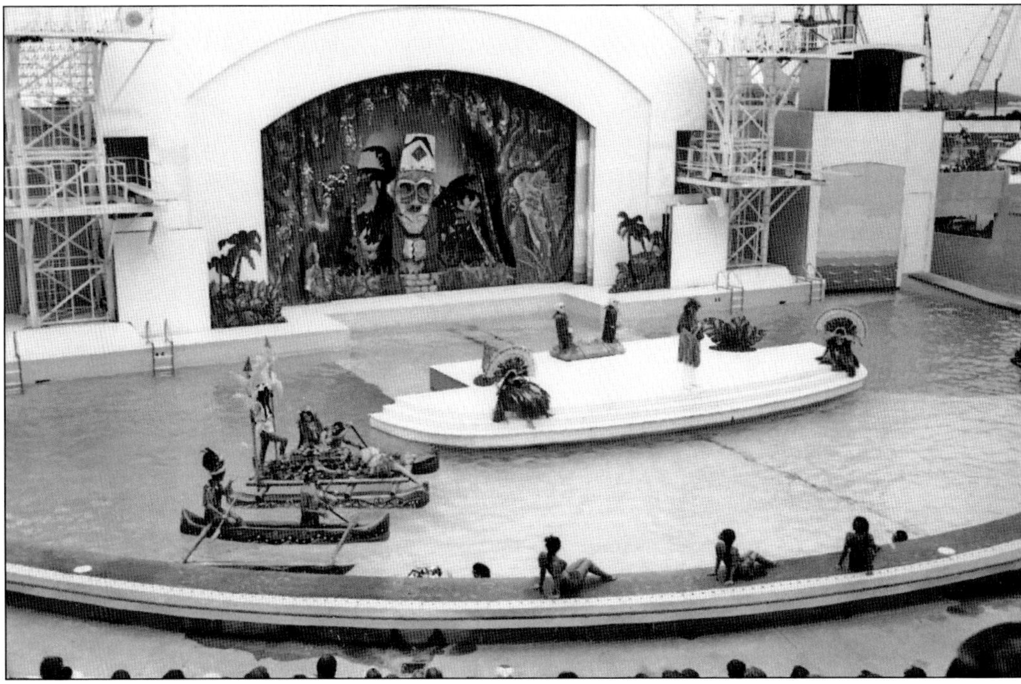

The 35-minute Aquacade show included an act from each decade from the 1920s through the 1980s, with songs, dances, and costumes from each era setting the mood. This segment from the 1940s used a moveable stage as part of a South Seas–themed routine.

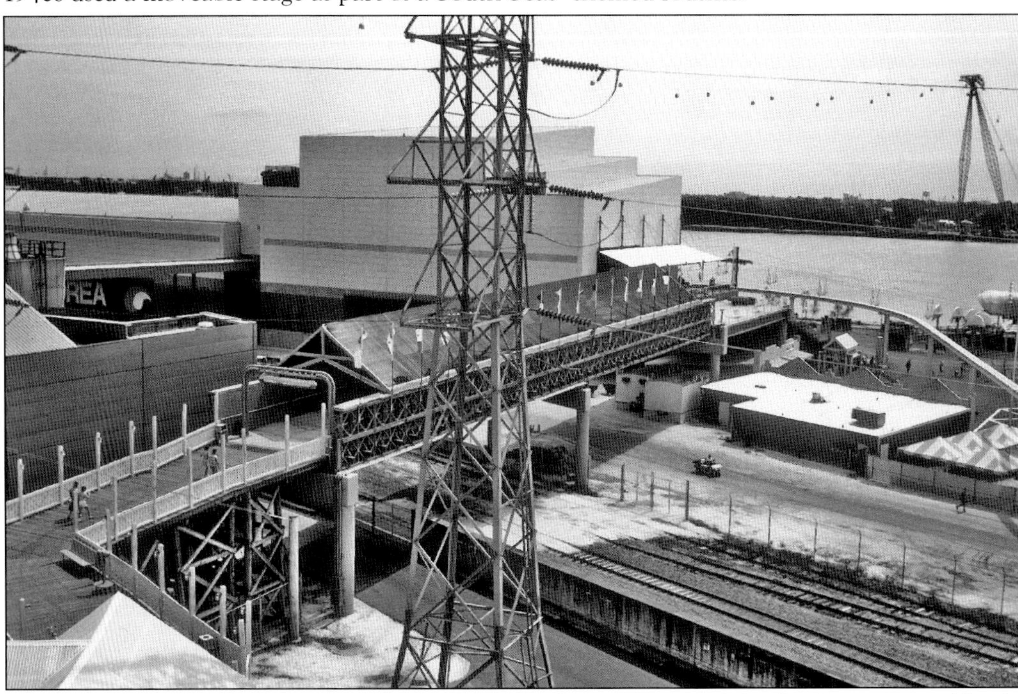

Bayou Plaza was linked to the International Riverfront section of the fair by a covered bridge that tried to block views of the railroad tracks that bisected the grounds. The gambit didn't work well, but few people lingered on the bridge due to the heat under its cover.

Six

The Great Hall

Between Bayou Plaza and Centennial Plaza is the Great Hall, a 15 acre covered facility that houses an intriguing variety of pavilions, exhibits, and services.

—1984 Annual Report

To say the world's fair was built around the city's convention center would be accurate in two ways. A look at any site plan or aerial photograph of the fair shows how the entire event was centered around one large structure right in the middle of the property. The fair and the convention center were also closely linked in another way. Both were launched in the 1970s to bring more tourists to New Orleans, and their births were tightly interwoven.

Plans were launched in 1978 to build a new facility to replace the Rivergate Convention Center, which was too small to attract the large gatherings the city needed. The new center would be more than three times the size of the old, with room to expand. Designed by the same architectural firm that was responsible for the fair's site plan, the new hall began construction in early 1981.

It had always been planned that the convention center's first tenant would be the world's fair, which leased the facility for the six months of the fair and named it the Great Hall. The main exhibits in the 380,000-square-foot exhibit halls were the states of Louisiana and Mississippi, both of which hoped to encourage visitors to spend more money there before heading home. These and the other exhibits were almost completed by the time a lease was finally signed. The terms called for $1.4 million in payments during the fair and an additional $2.5 million to come from the fair profits.

The fair never did make a profit, and the lease was never fully paid, but the convention center did benefit greatly from the fair. New Orleans finally had a facility that could attract major conventions, and following years of successful operation, the facility has been expanded twice since the days of the Great Hall. Now named the Ernest N. Morial Convention Center in honor of the mayor who championed its construction, the facility is the sixth largest convention center in the United States, with more than 1.1 million square feet of exhibit space.

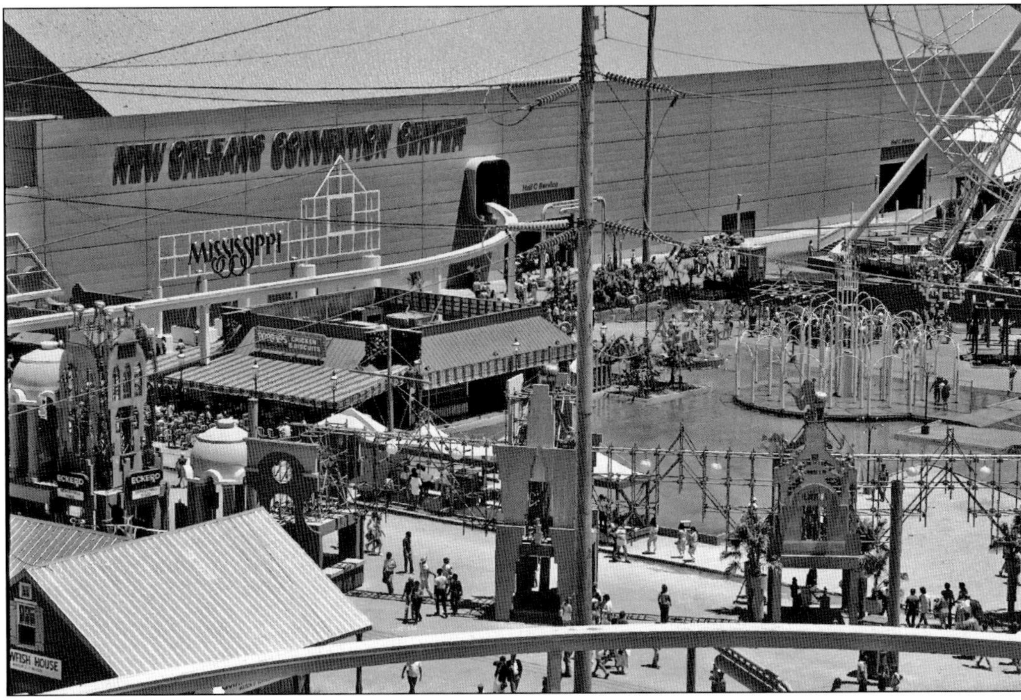

Little effort was made to disguise the Great Hall as anything other than the city's new convention center. While some banners were hung along parts of the wall to advertise the Louisiana exhibit inside, the rest of the building already had the name displayed on large signs intended for post-fair use. (Courtesy of the Ernest N. Morial Convention Center.)

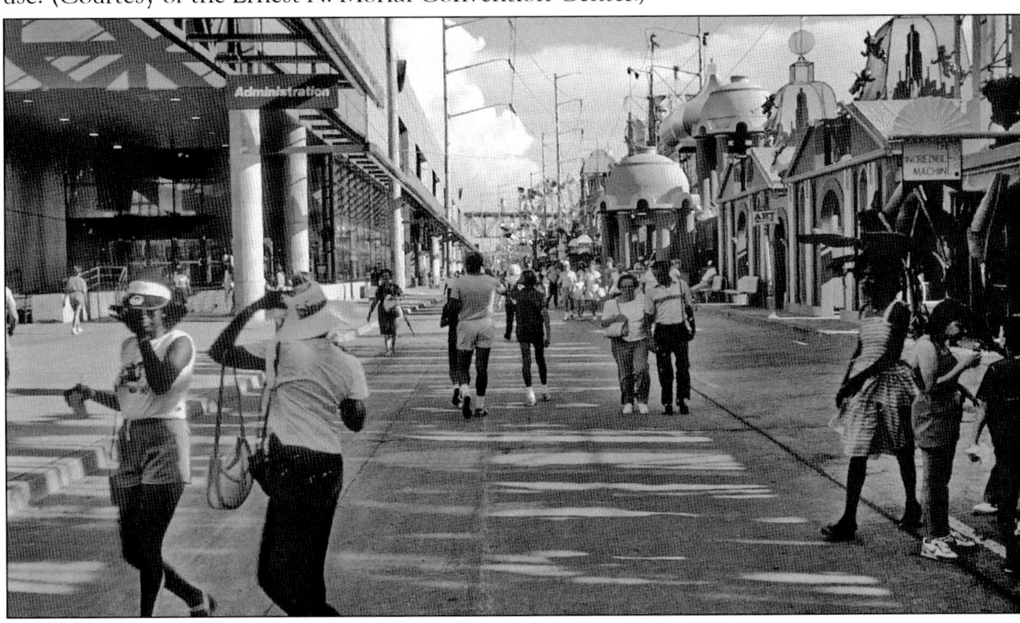

This street scene shows some of the incongruity that was the 1984 World's Fair. The angular shape of the new convention center building clashed with the fanciful shapes of the Wonderwall, but since the fair had been launched to help fund the center in the first place, there was little the organizers could have done to mask the building's utilitarian design.

The main attraction inside the Great Hall was the 130,000-square-foot Louisiana Journey. The air-conditioned boat ride through simulated swamps and bayous was a big hit during the hot summer months of the fair.

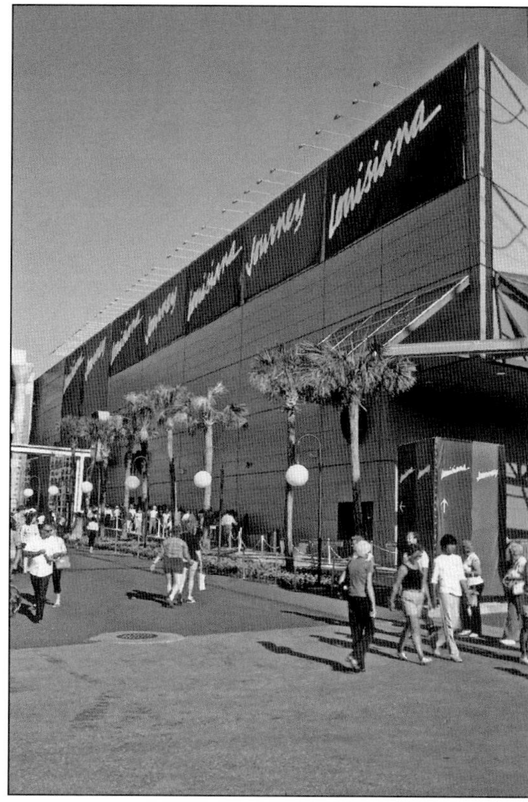

Long lines could often be seen waiting for the 14-minute ride, which used a variety of projection systems and special effects to tell the story of life today and in the past along the state's waterways. (Courtesy of the Ernest N. Morial Convention Center.)

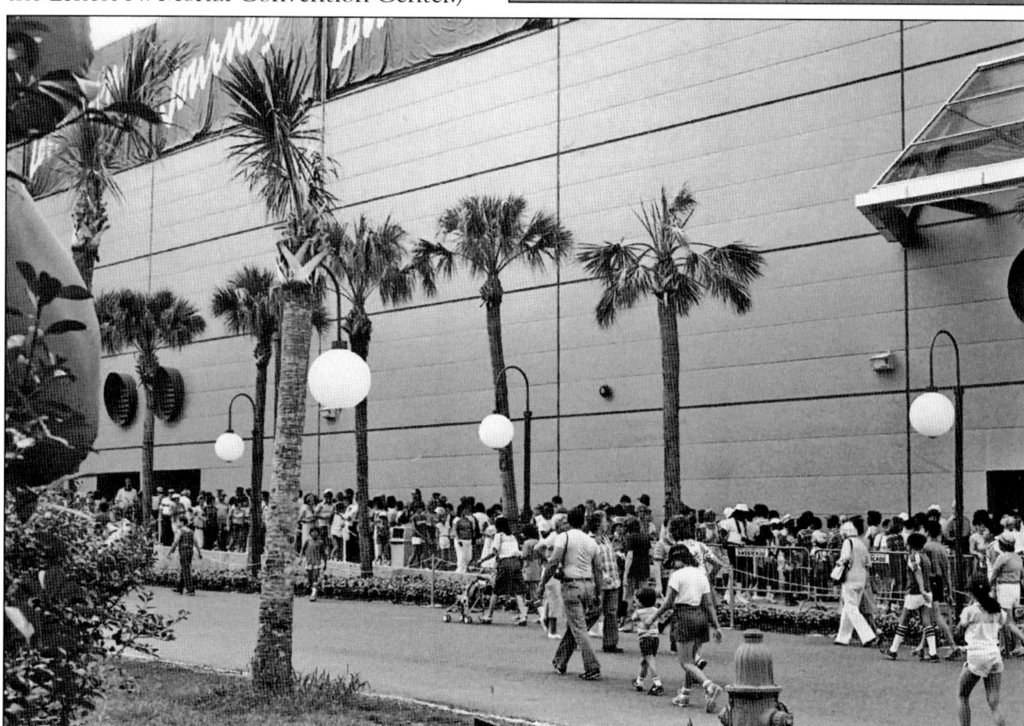

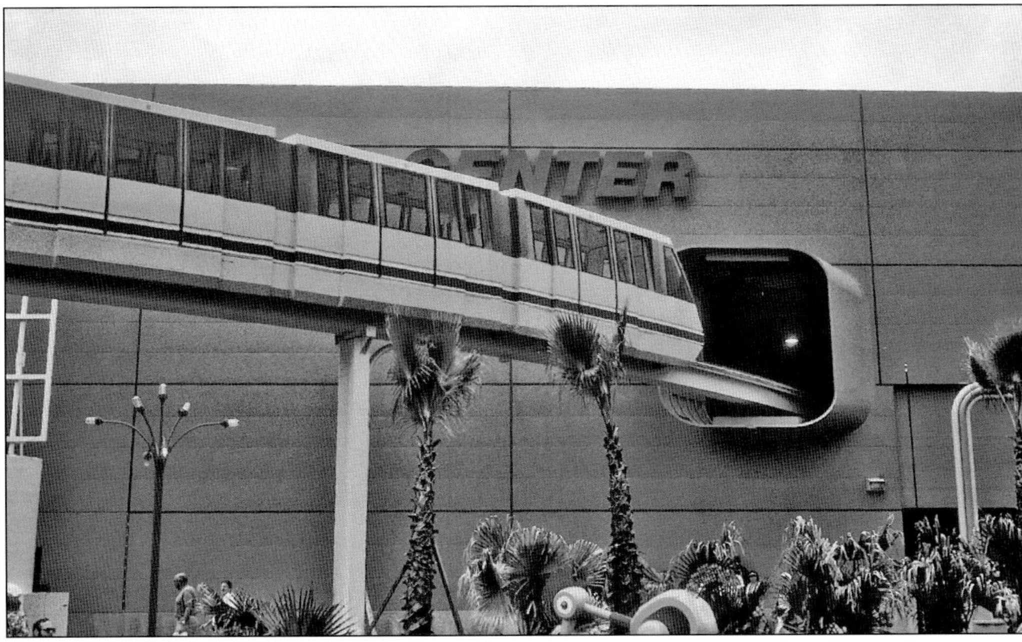

The Great Hall is probably unique in being the only convention center that featured a monorail system passing through it. While at least one travel writer covering the fair thought that the monorail might become a permanent mass-transit link to the convention center after the fair, it was never intended to be permanent.

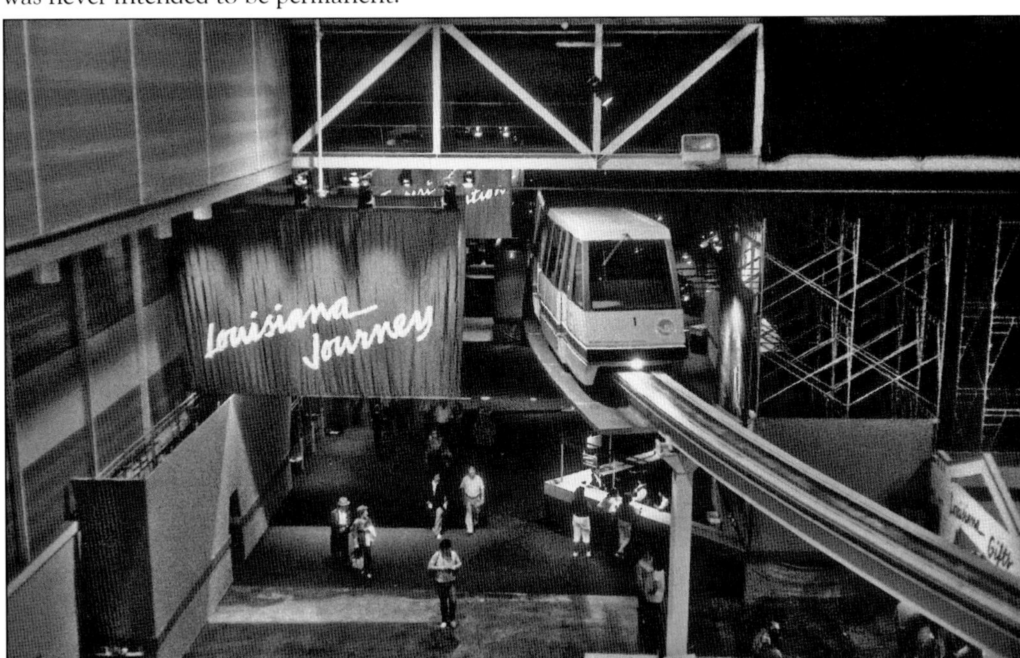

The monorail was strategically routed past the Louisiana Pavilion, giving riders a preview of the 1,800-foot-long boat ride. Twenty-four-passenger boats carried guests through a history of the state that began when the Earth was formed. Models, pictures, and elaborate set pieces traced the transformation from an agricultural to an industrial economy.

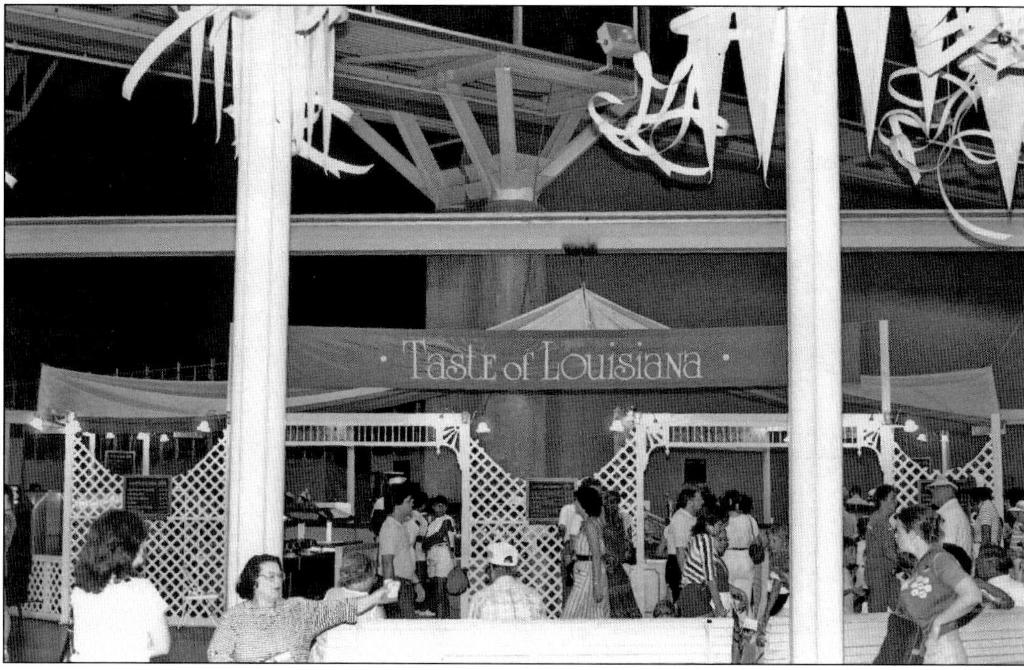

The pavilion also included Taste of Louisiana, which featured local food and drinks. The cafeteria-style facility served up a variety of Creole dishes, such as jambalaya and gumbo. For those not used to the spicy meals, there was also a welcome supply of cold beer and soft drinks. (Courtesy of the Ernest N. Morial Convention Center.)

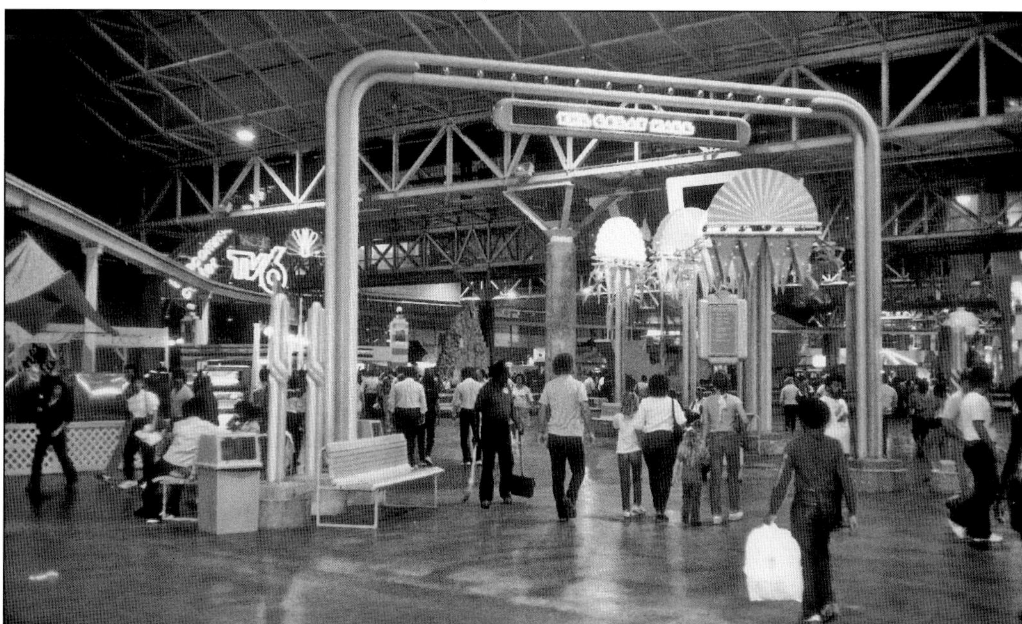

Visitors could also wander through the hall without having to wait for the Louisiana boat ride, for the Great Hall included a number of commercial, religious, and government exhibits. Colorful signage and other elements were used to break up the large expanse of concrete between the displays.

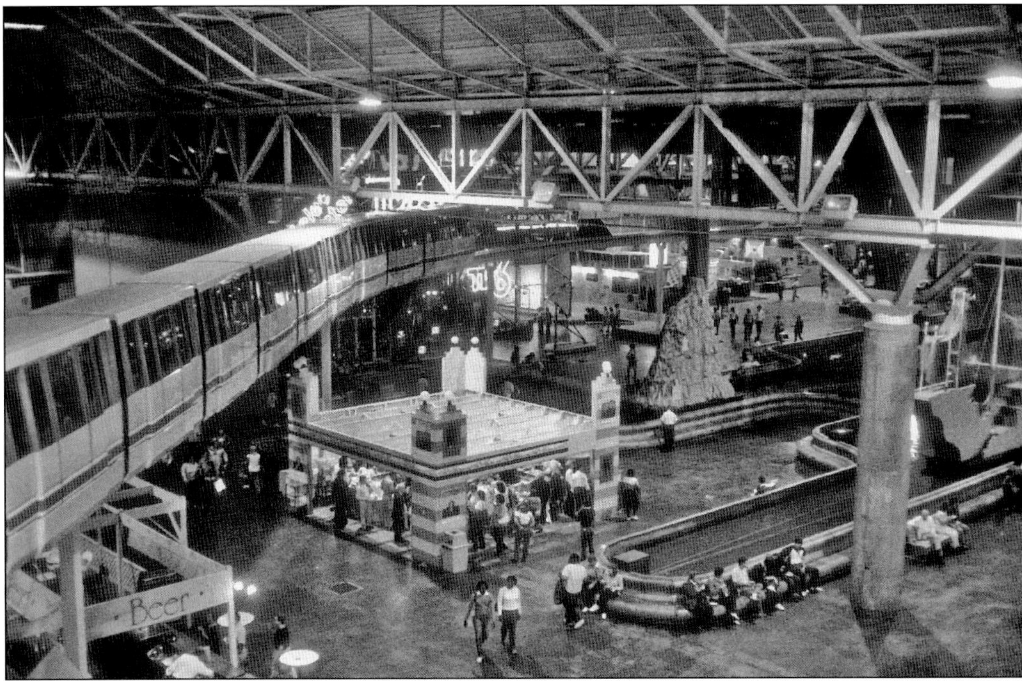

Here the monorail winds its way through the cavernous interior of the Great Hall. The square structure in the center of the picture was the largest souvenir stand at the fair, and next to it is one end of a canal system called the Water Course.

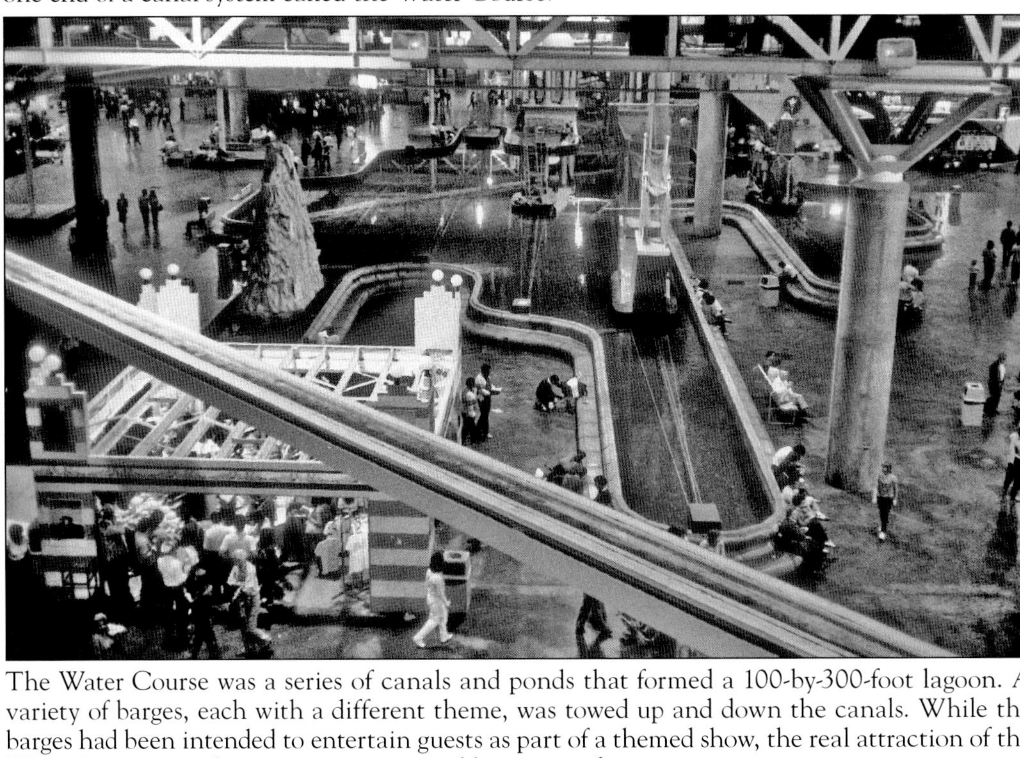

The Water Course was a series of canals and ponds that formed a 100-by-300-foot lagoon. A variety of barges, each with a different theme, was towed up and down the canals. While the barges had been intended to entertain guests as part of a themed show, the real attraction of the Water Course was that it was a way to cool hot, weary feet.

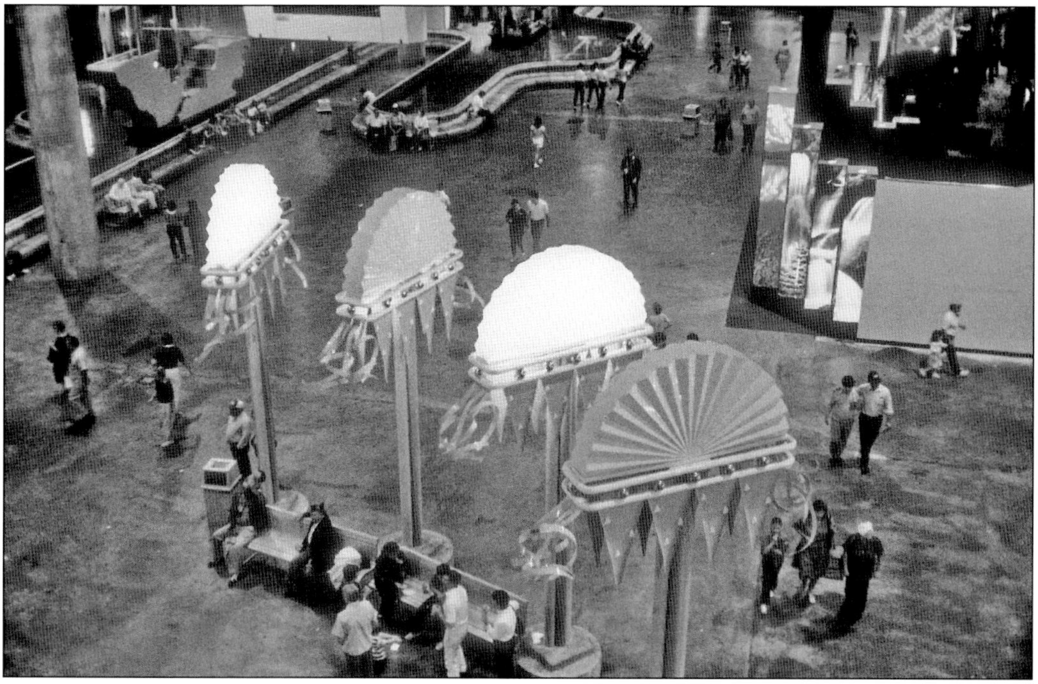

Another benefit of the Great Hall was that guests could sit inside, out of the sun, while pondering plans for the rest of the day. The benches scattered across the floor were often filled to capacity with hot and tired fairgoers, for there was not much shaded seating available outside.

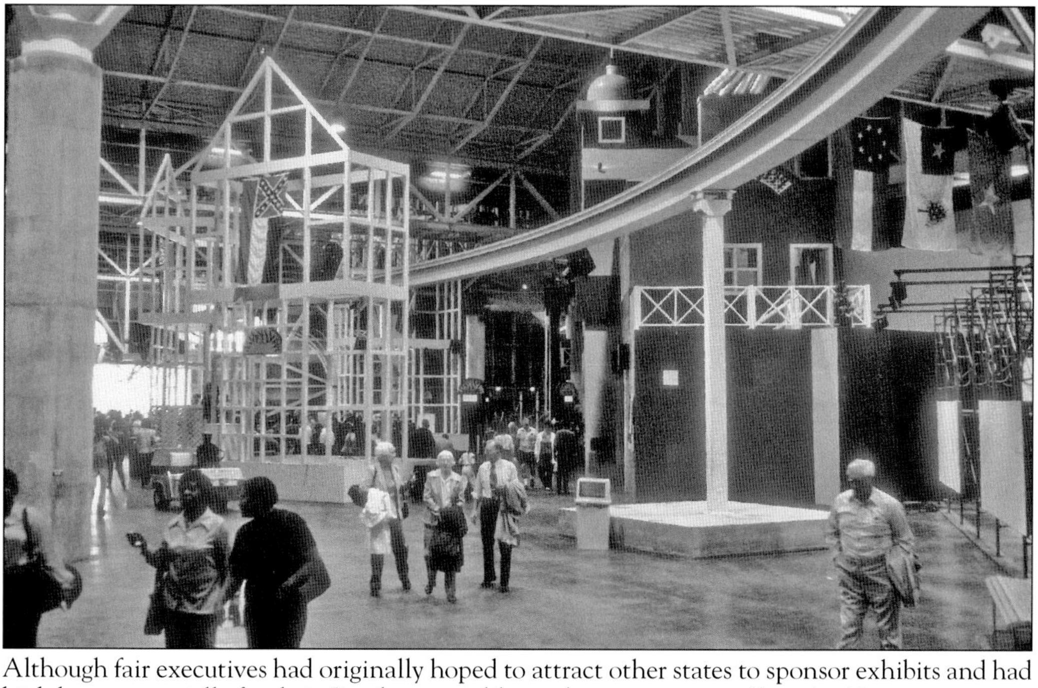

Although fair executives had originally hoped to attract other states to sponsor exhibits and had high hopes especially for their Southern neighbors, they were eventually only able to sign up one other state. Louisiana shared the Great Hall with the State of Mississippi Pavilion, which was the second largest exhibit inside the hall.

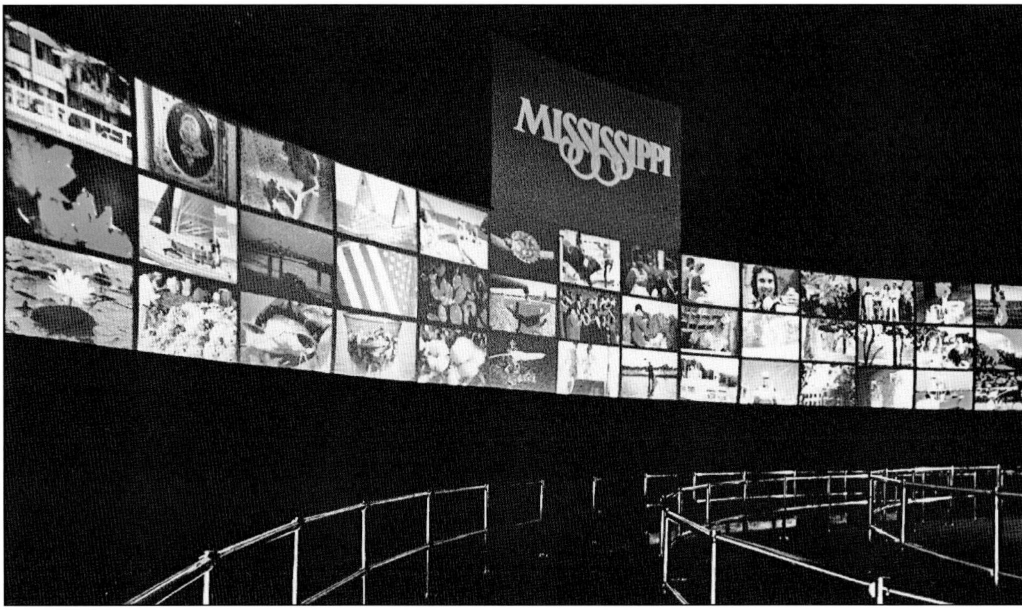

The Mississippi Pavilion featured 42 different slide projectors synchronized across a 140-foot-wide screen with narration by James Earl Jones. The show offered visitors a look at the state's past, present, and future, with an emphasis on the delights awaiting potential tourists. (Courtesy of Hap Owen.)

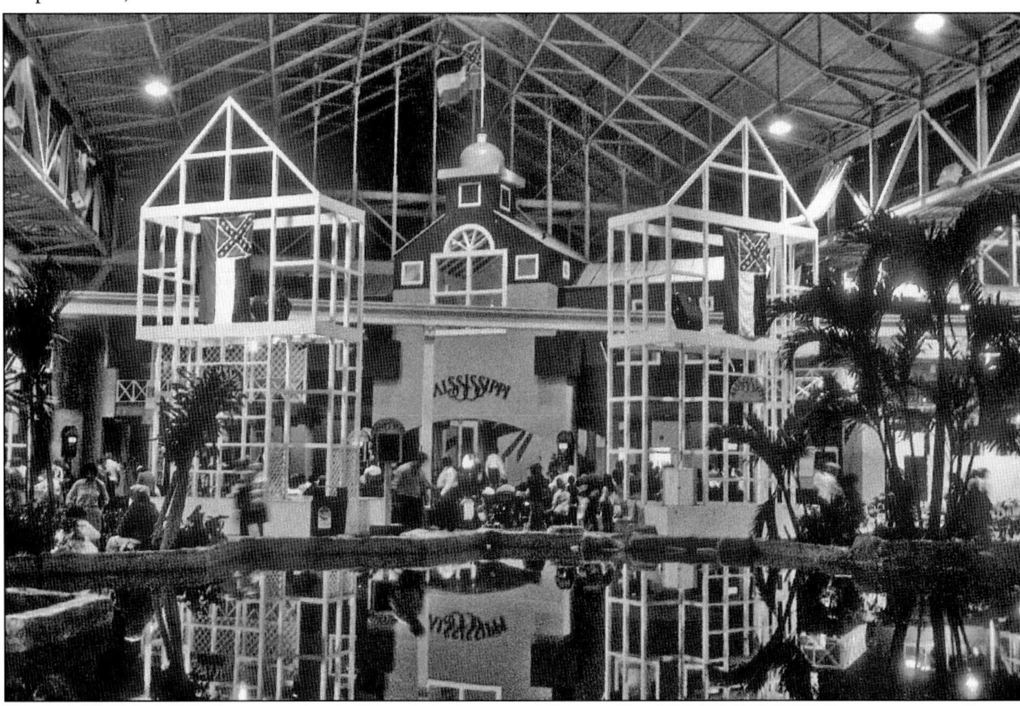

The entrance to the $2-million Mississippi display illustrates the vastness of the Great Hall and shows how it was well suited for later use as an exhibit center. The pavilion designers were able to make a relatively small structure look much larger through innovative features such as the building frameworks on either side of the main entrance.

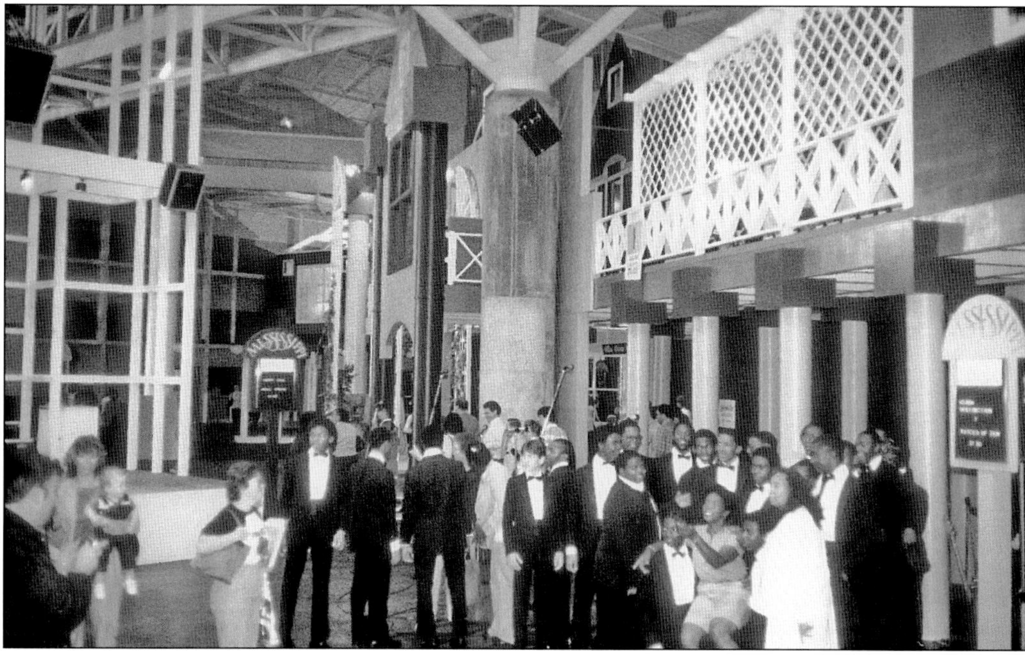

The Mississippi Pavilion included a section where groups from the state could entertain the passing crowds. The area was small but popular, as seen in this view of a church group leaving the stage as a boy's choir begins to sing. The many live performances across the fair gave it a local flavor that was popular with guests.

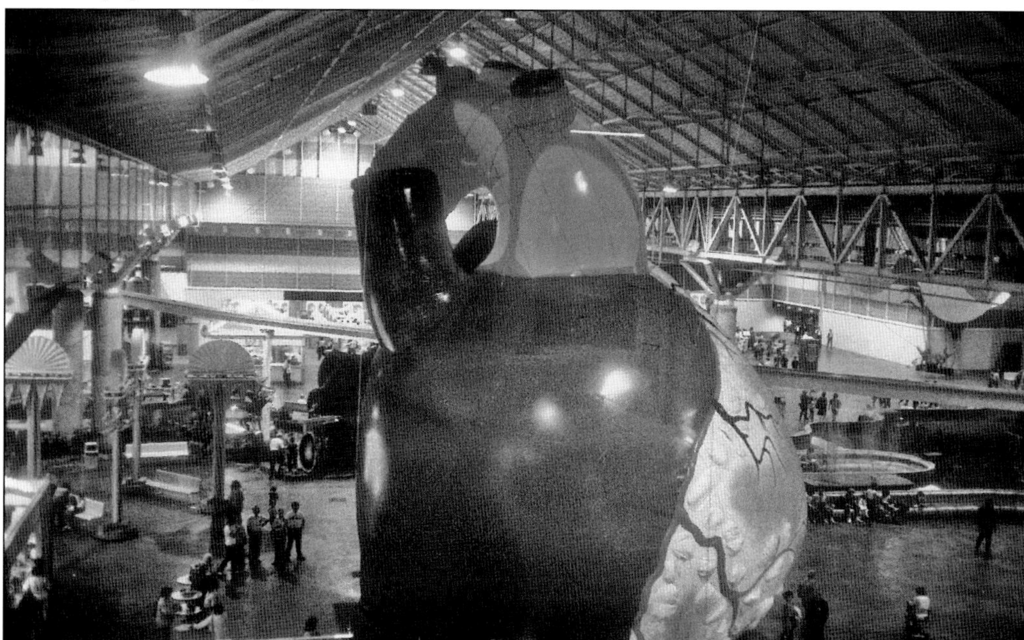

The Ochsner Pavilion, sponsored by New Orleans's Ochsner Medical Center, featured a 47-foot-high model of a human heart as the centerpiece of its exhibit about the cardiovascular system. Displays throughout the pavilion explained how the heart functions and the importance of diet and exercise in maintaining one's health.

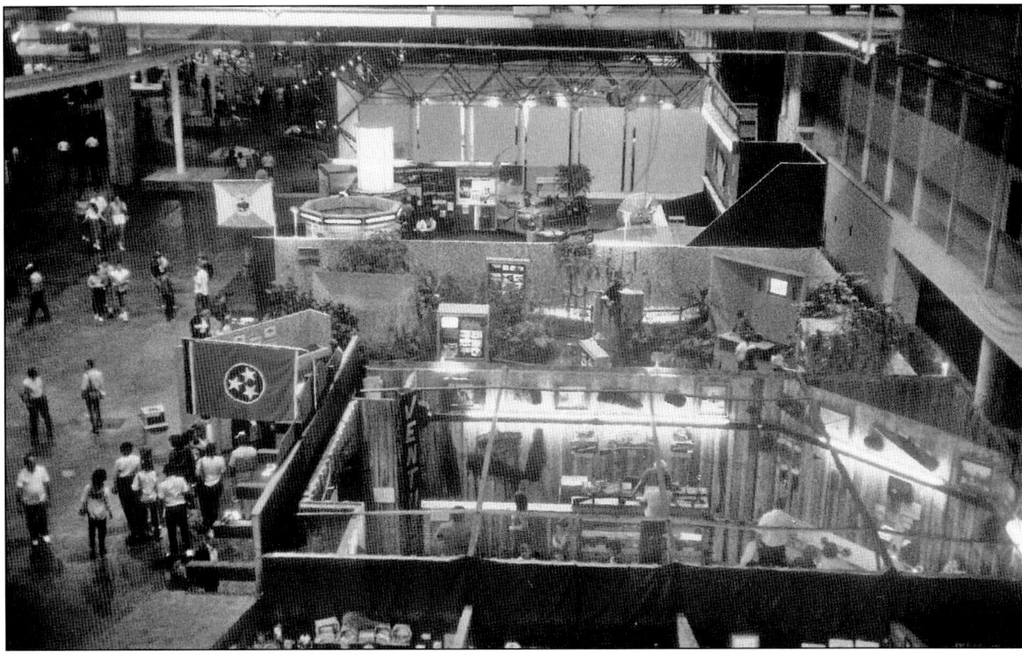

The fair's desire to attract exhibitors produced some unusual results, including one of the strangest exhibits at any world's fair. The booth in the foreground is Judge's Gun Shop, where visitors could peruse a selection of hunting rifles and other armaments. Hopefully visitors couldn't actually buy anything lethal there.

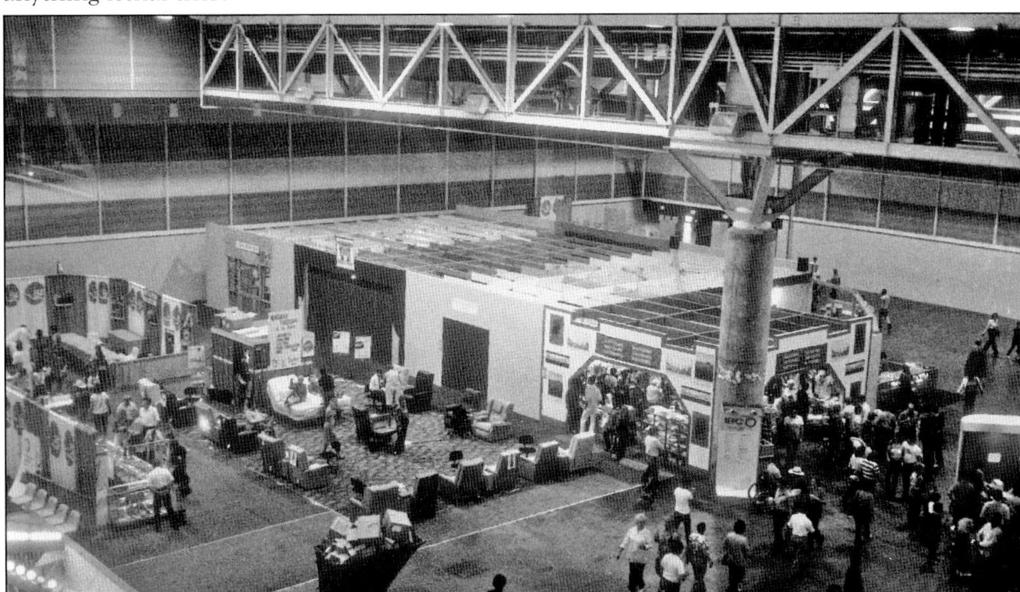

Desperate to fill the huge area as much as possible, fair officials continued to lower their standards when the hoped-for exhibits from major companies failed to materialize. This led to less-than-memorable displays, such as this collection of reclining chairs and footrests. Weary visitors may have enjoyed this section of the Great Hall, but most people probably did more shopping at the official souvenir stand next to it.

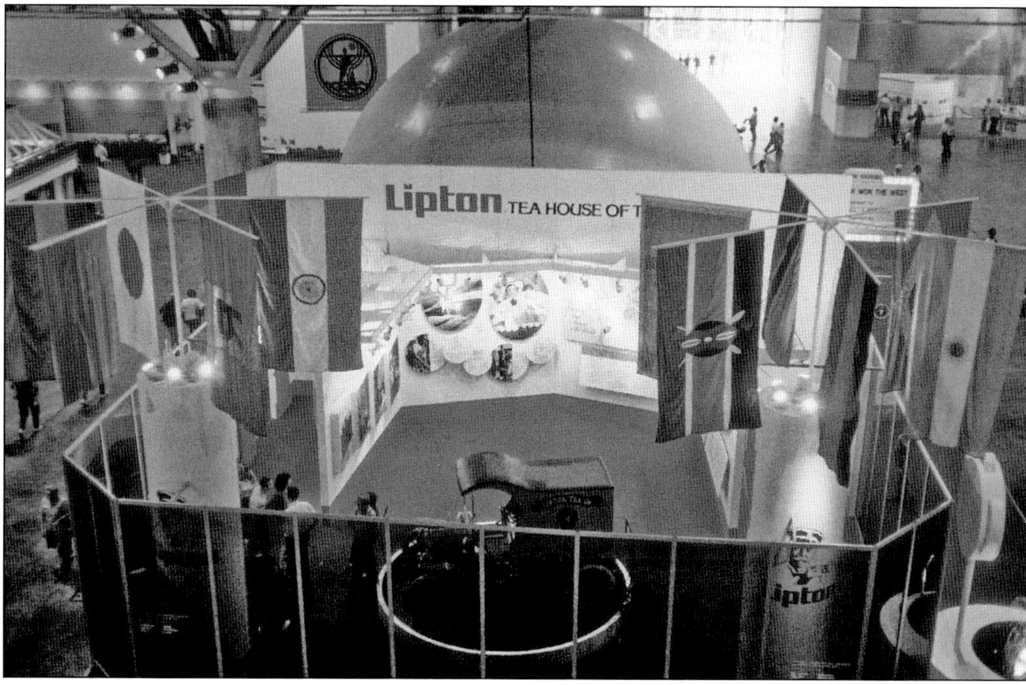

Lipton's Tea House of the World showed the path that tea takes from growing in remote plantations to when it is ready to drink. Facts about the many different types of tea surrounded the centerpiece of the display, a 1912 Lipton delivery truck said to have been used by Sir Thomas Lipton himself.

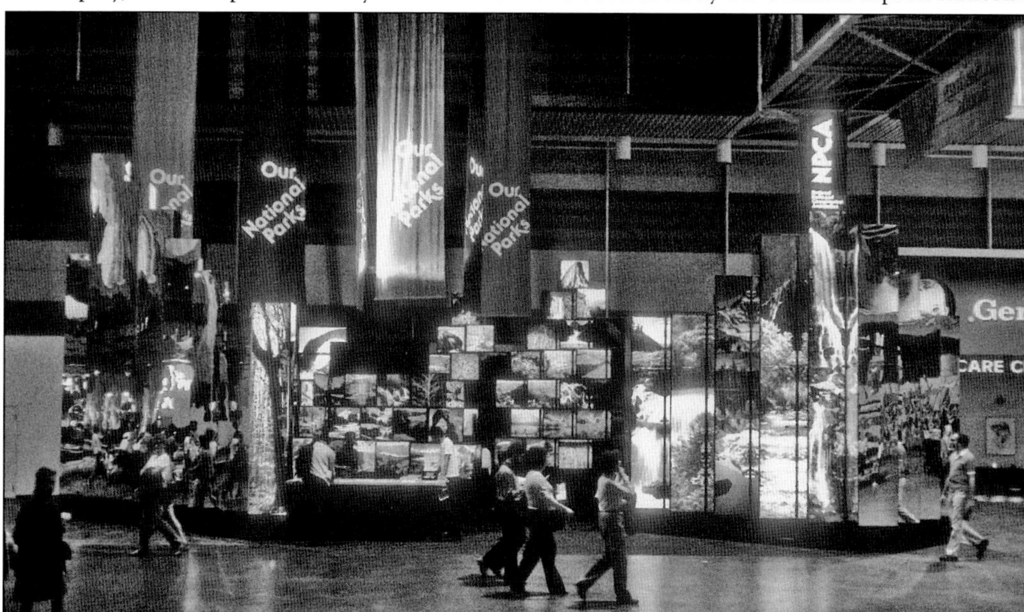

The National Parks and Conservation Association showcased many of the country's natural and historic wonders in this colorful display that featured illuminated photographs. A gift shop offered Native American products and other souvenirs from the national park system. The Gerber Baby Changing Station, seen just to the right of the park service exhibit, was also the fair's center for lost children.

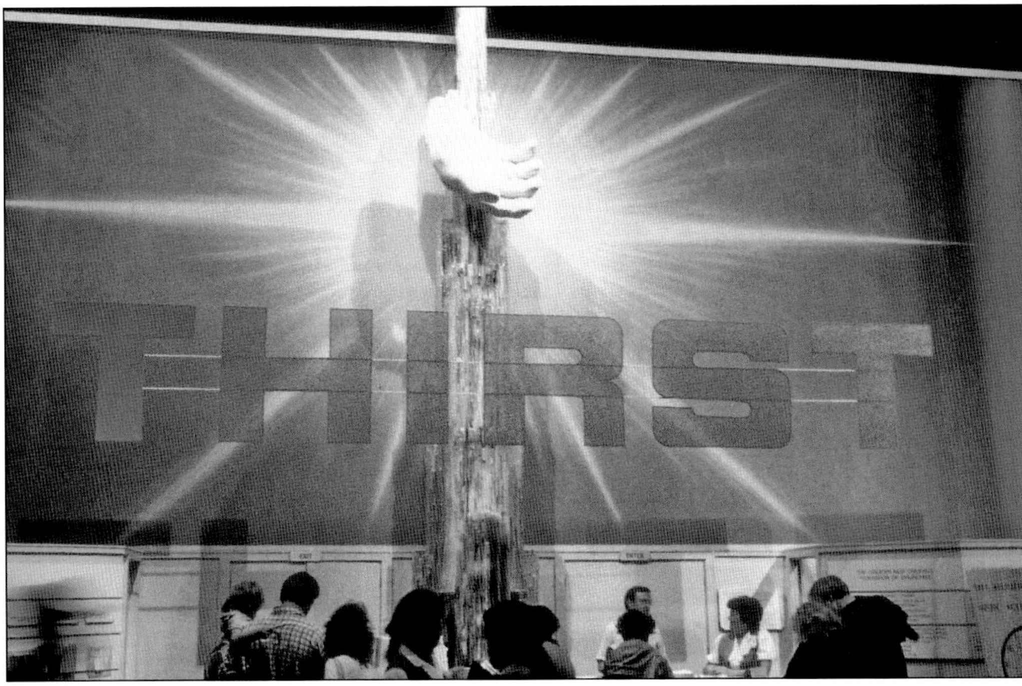

The Louisiana World's Fair Ministries Pavilion, also known as the Christian Pavilion, tied in with the fair's river theme by describing how people "thirst" for religious elements in their lives. In addition to a theater show, the pavilion included a display of rare bibles and other religious documents.

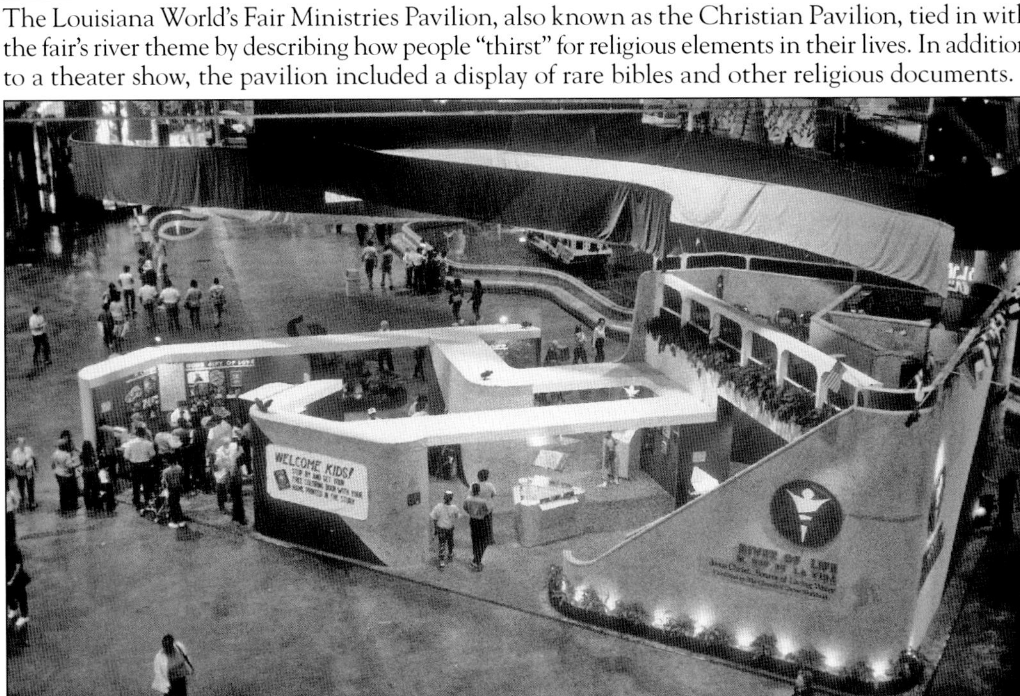

"River of Life" was the theme for the Church of Christ Pavilion. A variety of religious exhibits were on display with different sections for adults and children. Computer screens offered short films on religious subjects, and volunteers were nearby to answer questions or to hand out literature. Another area was available for those who wanted to have a more personal discussion about their religious options.

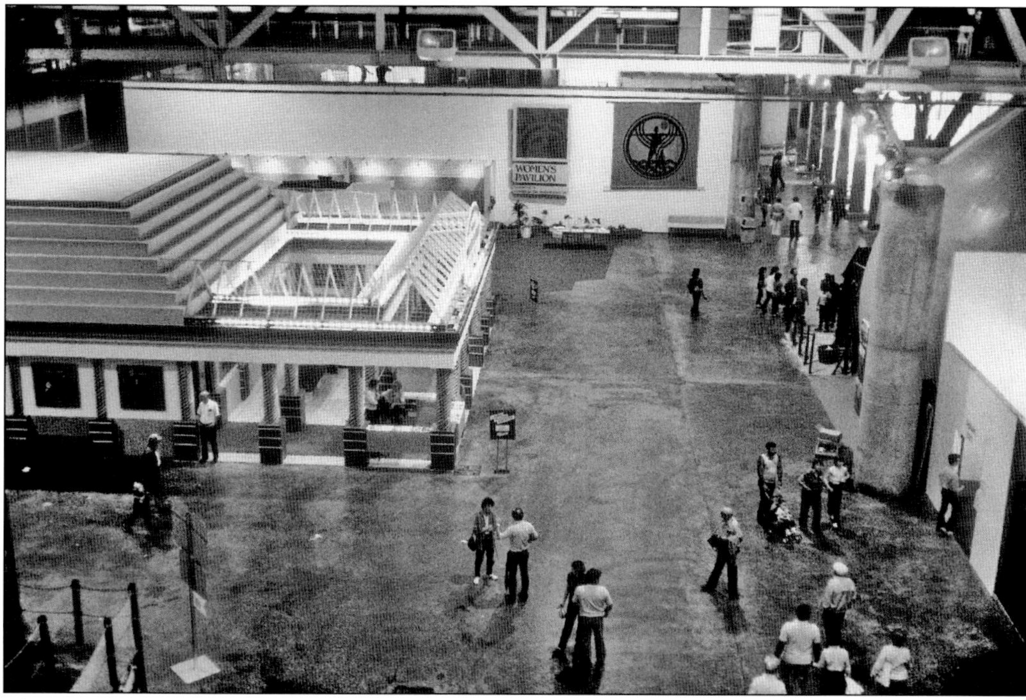

The Living Water Theatre, on the left, was sponsored by the Seventh Day Adventist Church. A 13-minute show that journeyed back to the creation of the Earth explained the pavilion's theme, "Christ, the Source of Living Water." Biblical quotes about water were also displayed throughout the pavilion. The Women's Pavilion, on the right, was originally planned as a more appealing geodesic dome but was reduced to a simpler display area by the time it opened.

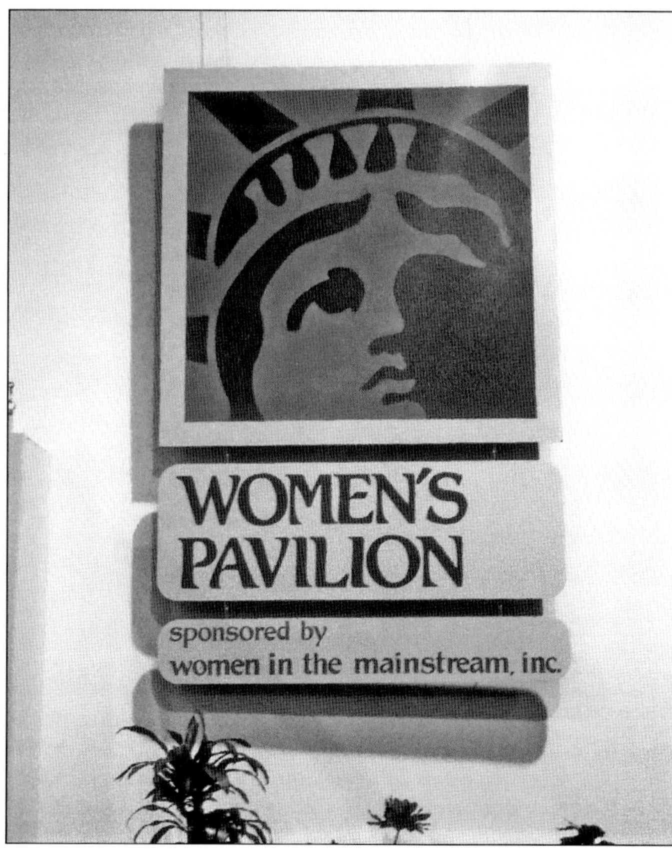

The Women's Pavilion saluted contributions by women from around the world during the past 100 years. The pavilion also included an art exhibit of award-winning works by American women.

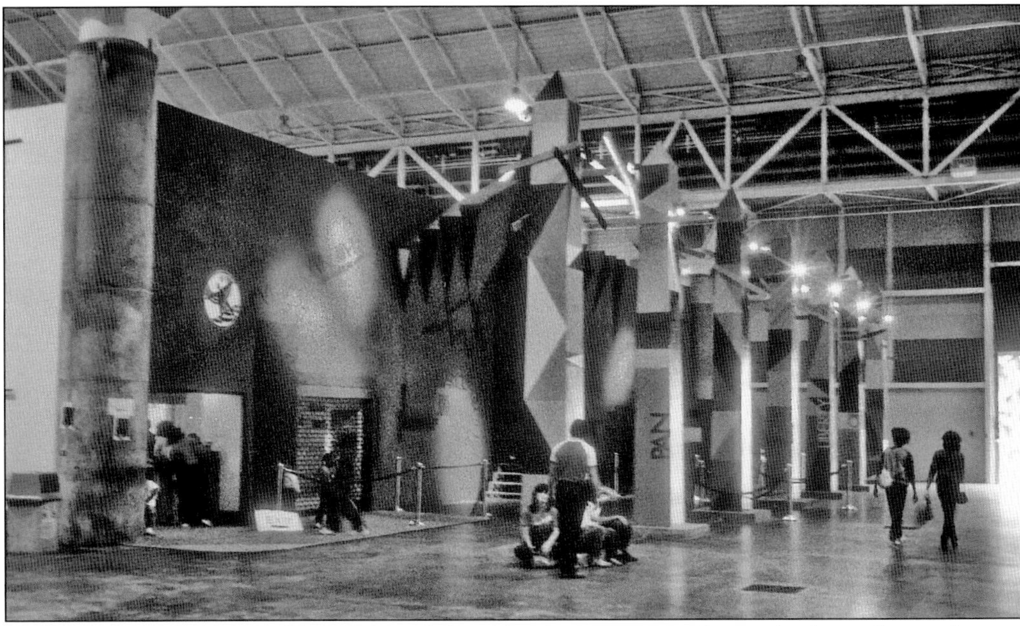

"I've Known Rivers" was the message of the Afro-American Pavilion, which used several water-themed segments to tell the past and current challenges faced by African Americans in America. For example, the trip across the ocean to slavery was told in a short presentation that led to an examination of life on a plantation. Other exhibits saluted scientific and cultural contributions by African Americans.

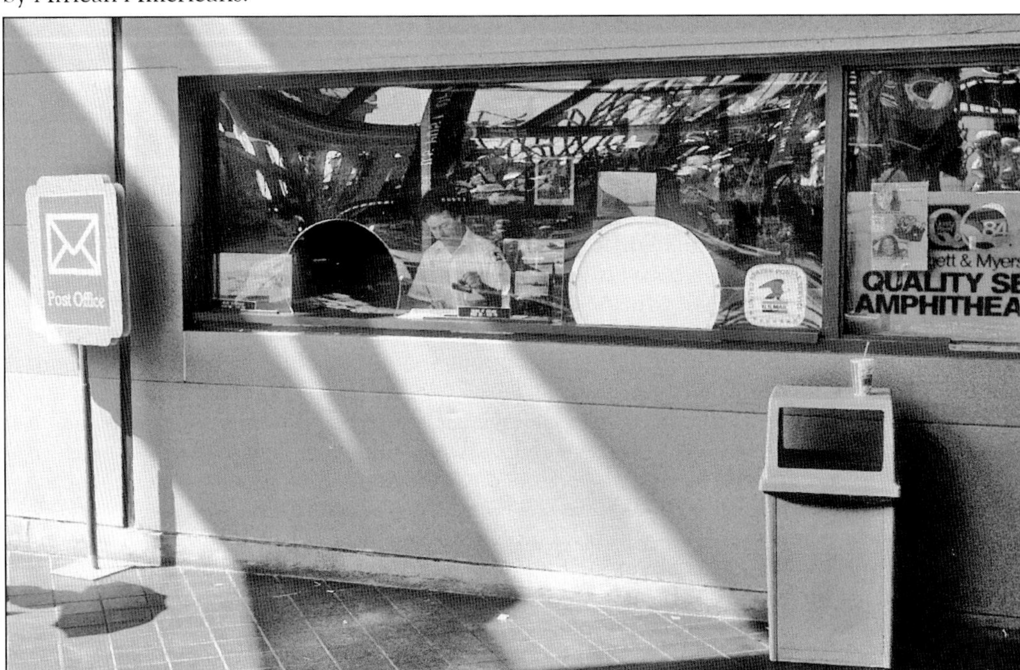

Perhaps the loneliest job at the fair, the U.S. Postal Service had a full-service station at one of the entrances to the Great Hall. Other than selling souvenir envelopes and stamps for postcards, it is hard to imagine how the clerks passed their time.

Seven

INTERNATIONAL RIVERFRONT

The theme of the 1984 World's Fair, "The World of Rivers: Fresh Water as a Source of Life," evident throughout the site, takes on global impact on the International Riverfront. Each country explores and showcases its history, understanding, and technological advances in the field of water-related issues.

—1984 Annual Report

The key to any world's fair is international participation. Without it, a fair would be little more than a trade show or an overly glorified county fair. With it, the fair becomes an opportunity to see people and displays from parts of the world one may never be able to see in person. It was with this thought in mind that a significant portion of the fair site was set aside for foreign exhibitions.

The section chosen was one of the most desirable at the fair, stretching along the Mississippi River and covering several acres once occupied by little-used docks. The river locale made this part of the fair easy to spot from passing boats and the nearby highway, thereby keeping the event in the mind of those living in or visiting the area. Aptly named the International Riverfront, the area consisted of a 360,000-square-foot building that housed the majority of the pavilions and the United States Pavilion, located across a courtyard. The international pavilion building was a permanent structure, as part of an agreement with the dock owners. The International Riverfront also included an amphitheater, which was one of the most popular attractions at the fair; a monorail station; and in a major coup for the fair, part of NASA's Space Shuttle fleet.

The area almost didn't have any international exhibits. Potential participants balked at the costs involved in appearing at Knoxville's 1982 World's Fair, this fair, and also at Vancouver's Expo 86. Worried fair officials turned for help to the state's senators, who in turn contacted Pres. Ronald Reagan. His backing of the fair resulted in the U.S. State Department joining the effort, and the fair got its international tenants.

After the fair, the United States Pavilion and amphitheater were removed, and the remaining buildings were converted into a cruise ship terminal and shopping mall. Like the convention center, the former fair buildings have become an important part of today's New Orleans.

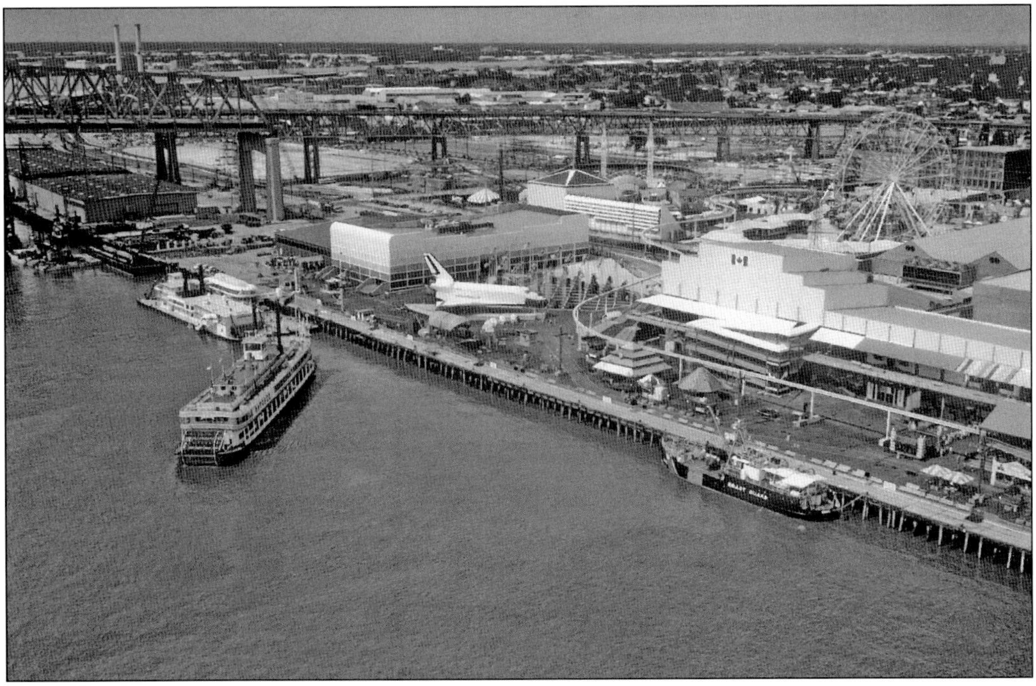

The southern end of the International Riverfront area is seen from the MART ride. A variety of vessels were moored here at different times during the fair, many offering free tours. The riverboat approaching the dock provided trips along the river for an additional fee.

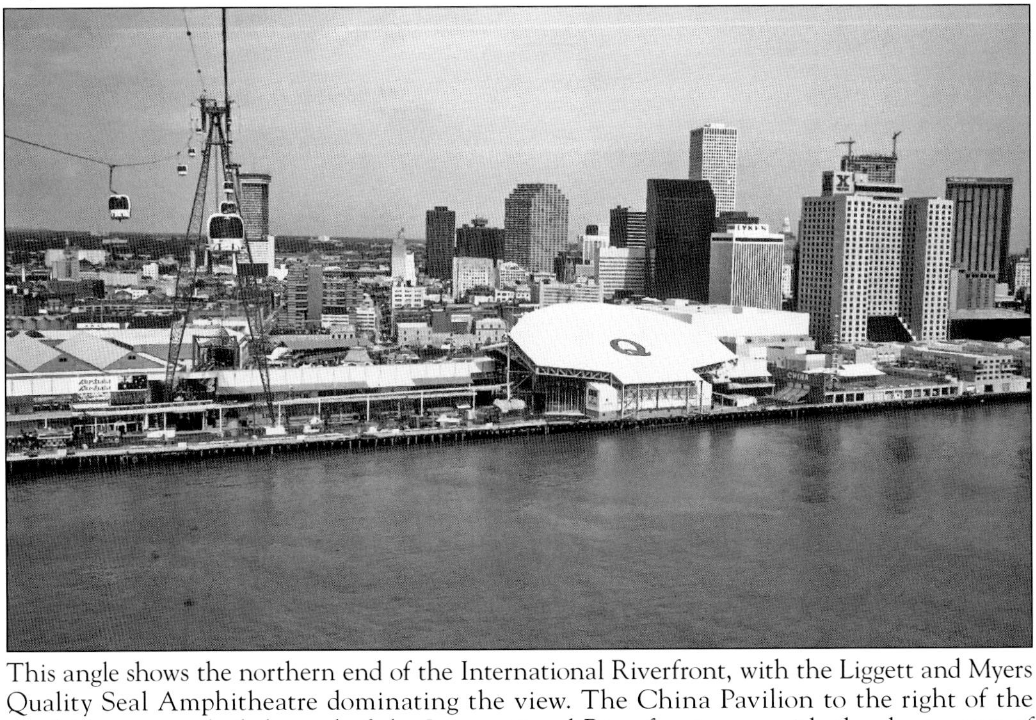

This angle shows the northern end of the International Riverfront, with the Liggett and Myers Quality Seal Amphitheatre dominating the view. The China Pavilion to the right of the amphitheater marked the end of the International Riverfront area, with the skyscrapers of modern New Orleans peeking up just outside the fair property.

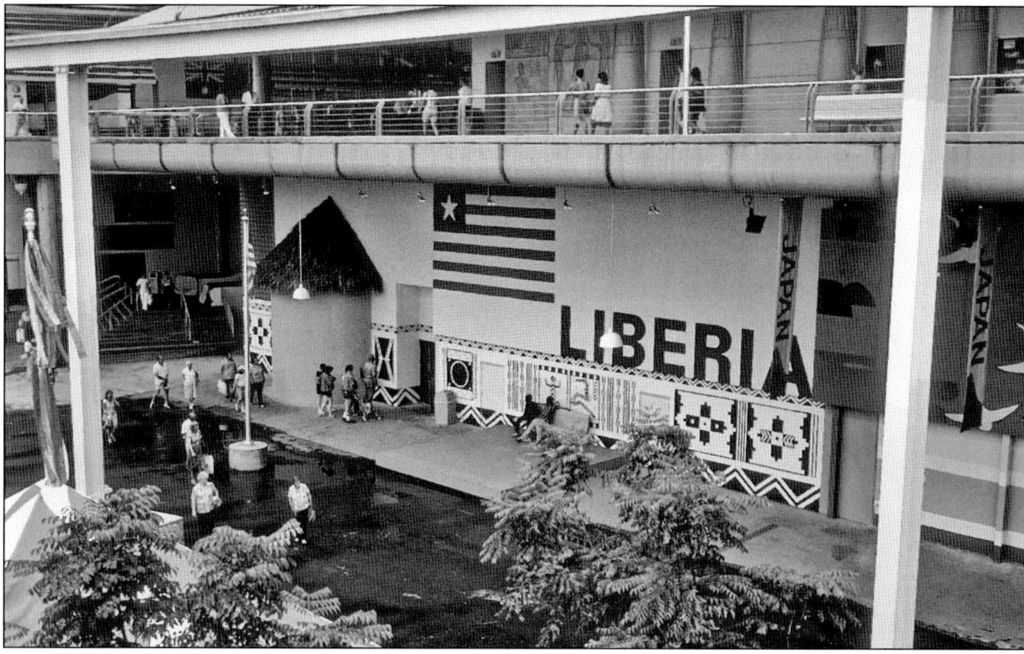

The majority of the International Riverfront area was a long two-story structure divided into different-sized segments to meet the requirements of each country. Some countries had two-story pavilions, while others were more modest. This view of the Liberia Pavilion entrance was taken from the Sky Transpo aerial tram, which ran along most of the riverfront.

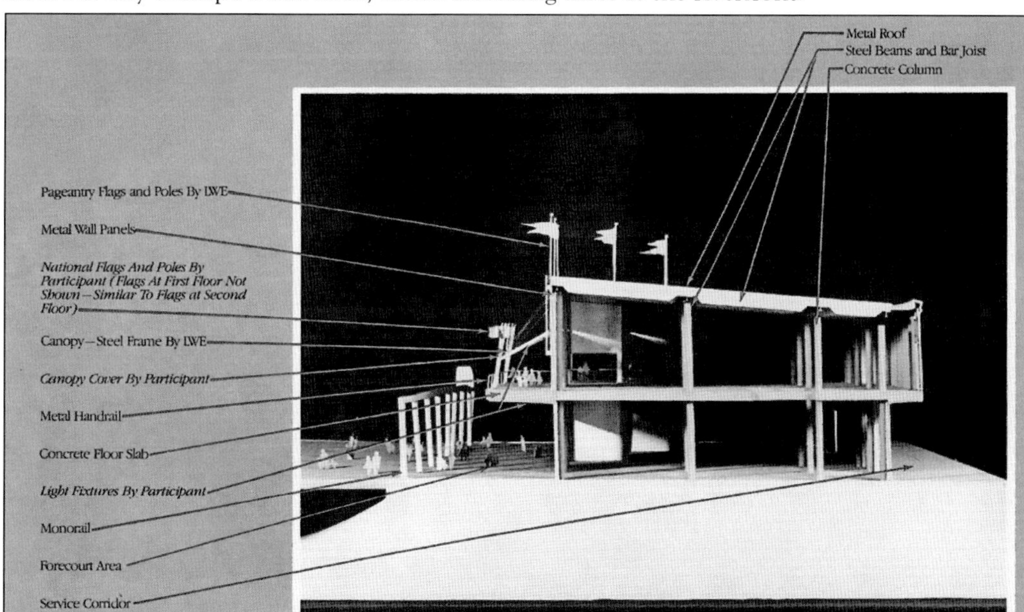

The master plan for the international pavilion gave the designers of the individual country exhibits very little leeway in their own designs. The structure, designed for post-fair use as a shipping facility, was essentially a long warehouse-style building. As this page from the official Design Guidelines book illustrates, the exhibits all had to fit within the fixed walls and support beams required for the building's future use.

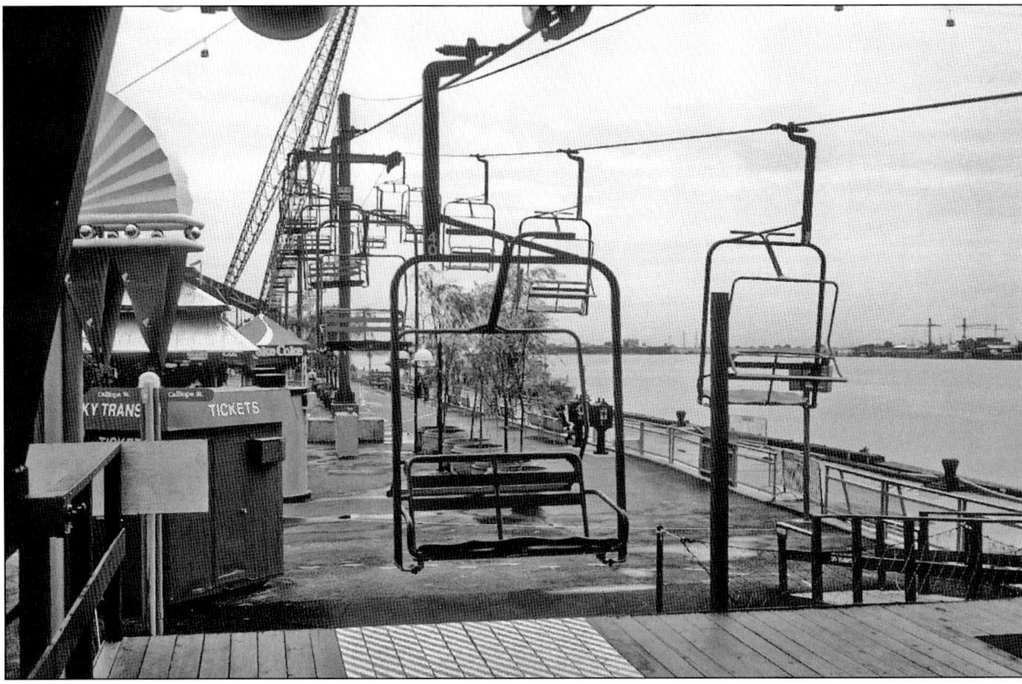

The Sky Transpo was basically a renamed aerial lift system that would more commonly be seen at ski resorts. It carried riders along the Mississippi riverfront for the length of the exhibit area. The higher gondolas seen in the background are from the MART system, which actually crossed the river.

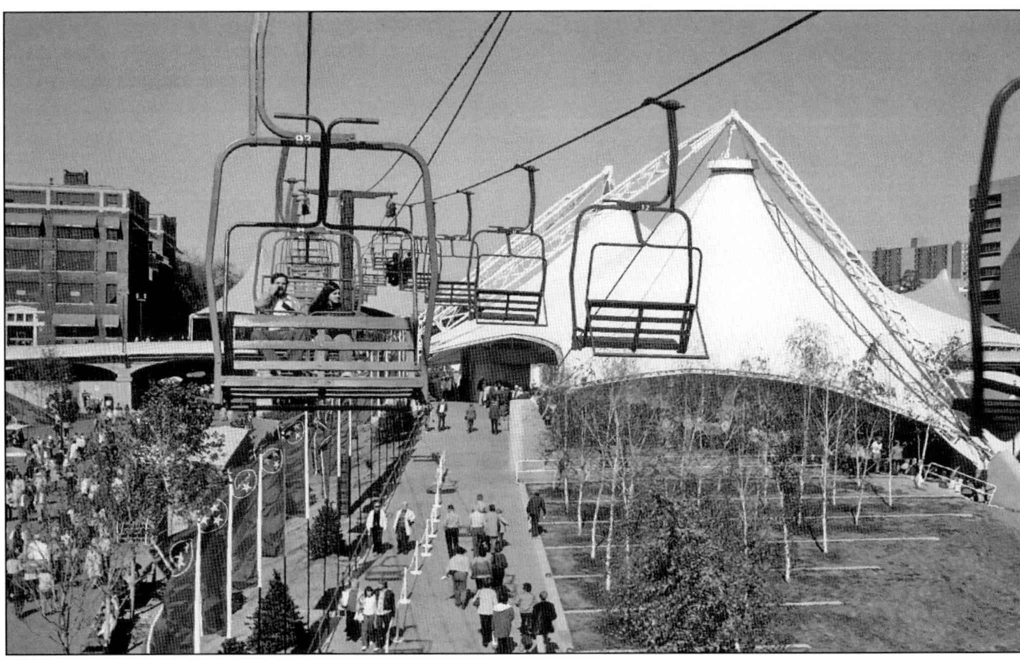

Unlike MART, Sky Transpo was not new to the fair, as seen in this view from the 1982 World's Fair in Knoxville. The ride was popular at both fairs, and it was often crowded in New Orleans at night, as it afforded excellent views of the fireworks show on the river.

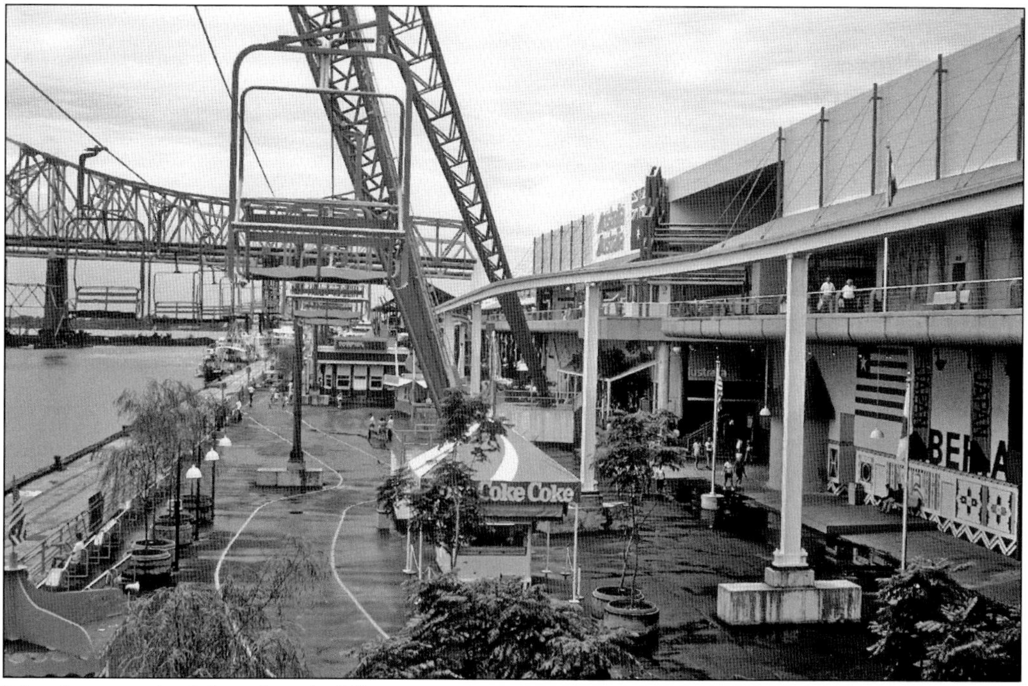

The designers had their hands full with this section of the fair, where they had to somehow fit in the supports for the Sky Transpo, MART, and the monorail while still providing enough room for visitors to stroll through the area.

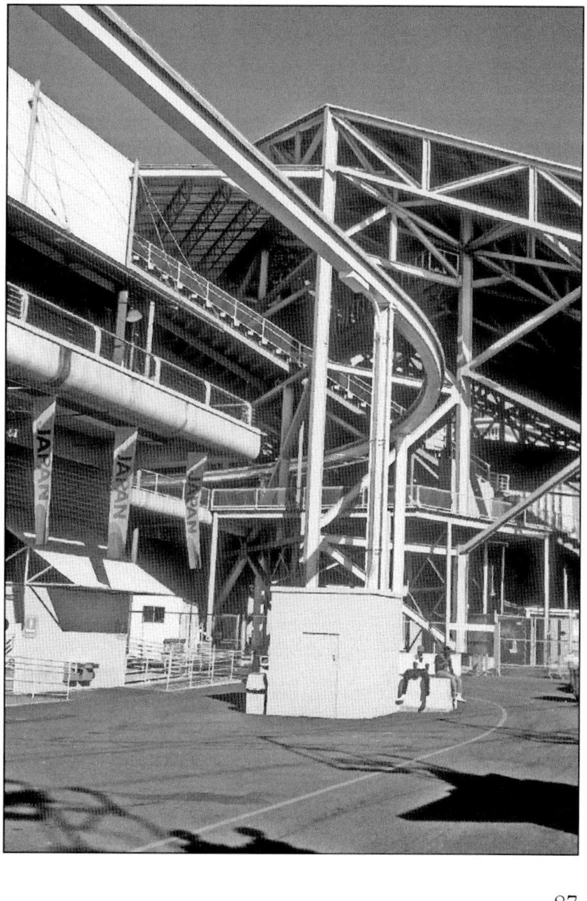

Things got even more complicated at the northern end of the main riverfront building, for the monorail had to make a tight left turn back to the rest of the fair and wind through the supports of the amphitheater.

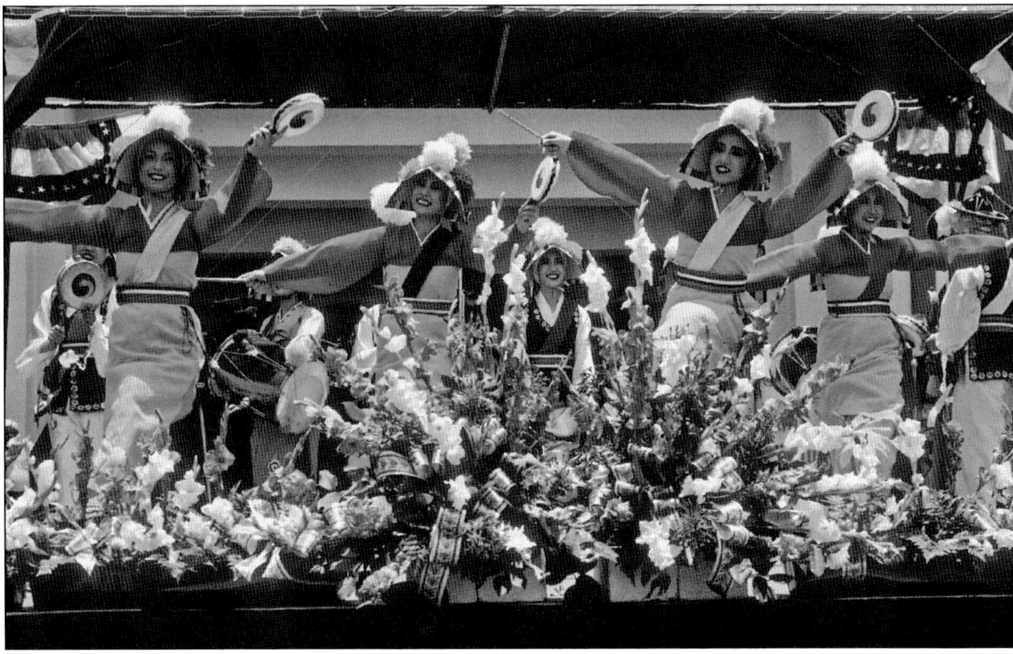

Part of the fun of every world's fair is learning about different cultures. Many of the international pavilions featured performers such as these entertaining outside the Korea Pavilion. Hopefully fair guests would be intrigued by these street shows, come inside for a more detailed look at the host country, and perhaps eventually travel there. (Courtesy of Mitchel Osborne.)

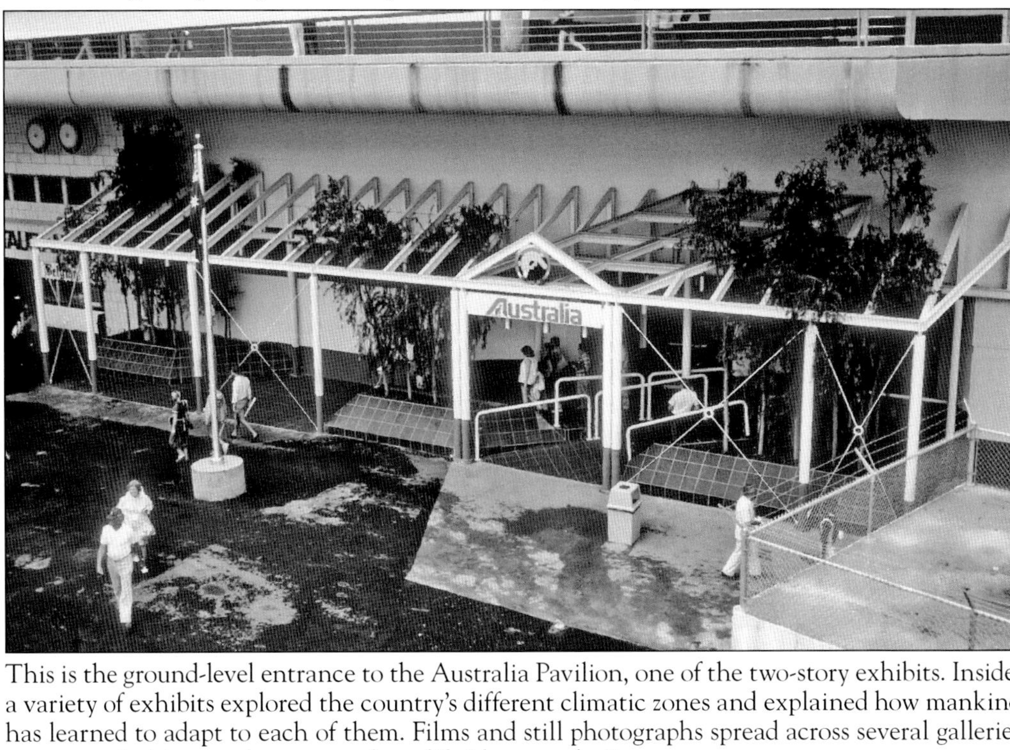

This is the ground-level entrance to the Australia Pavilion, one of the two-story exhibits. Inside, a variety of exhibits explored the country's different climatic zones and explained how mankind has learned to adapt to each of them. Films and still photographs spread across several galleries encouraged visitors to lean more about life "down under."

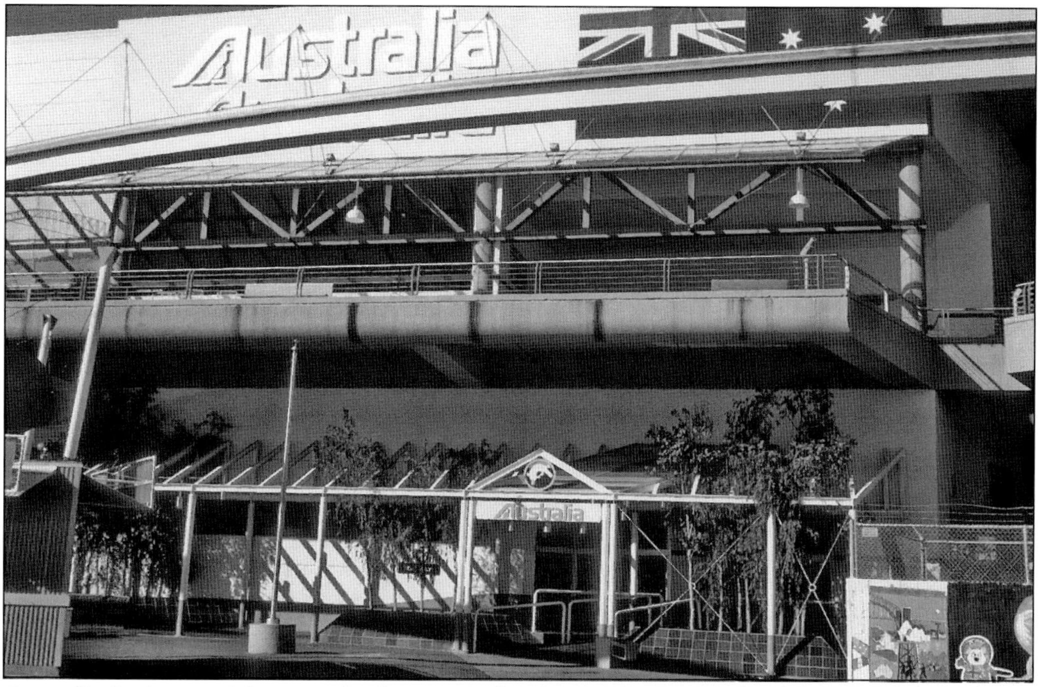

Australia had a special reason for having such a large presence at the fair: one of the exhibits inside showed plans and models from World Expo 88, the first world's fair to be held in Australia in more than 100 years. In an interesting twist, the Brisbane-based fair would later include a restaurant with a New Orleans theme.

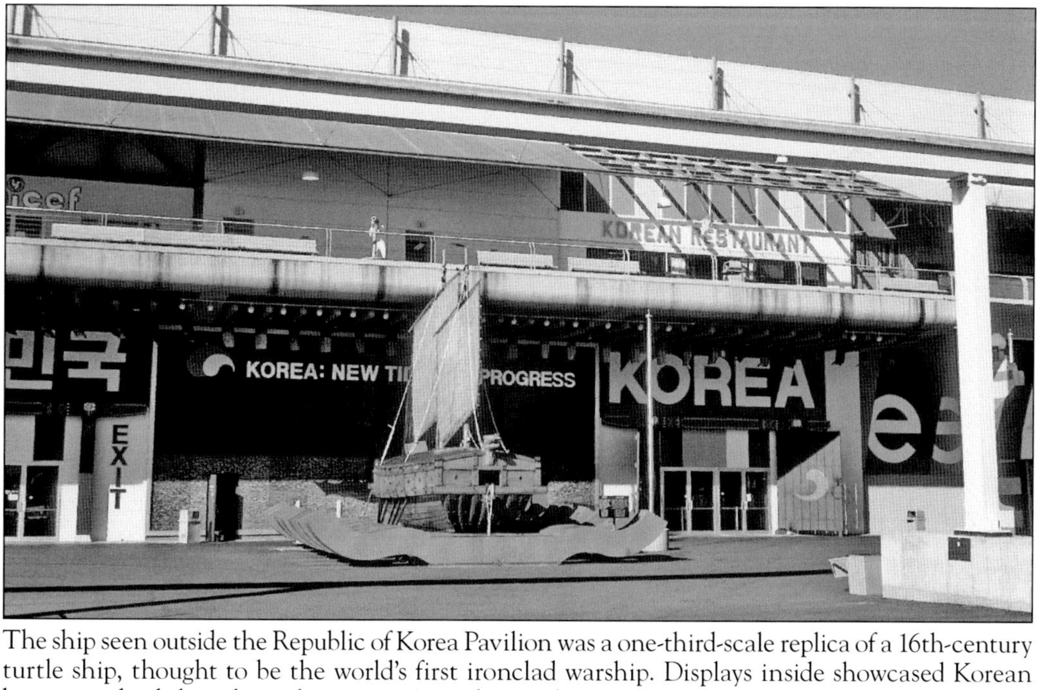

The ship seen outside the Republic of Korea Pavilion was a one-third-scale replica of a 16th-century turtle ship, thought to be the world's first ironclad warship. Displays inside showcased Korean history and exhibits about the country's modern industries. Upstairs, visitors could enjoy Korean food and products in the restaurant and gift shop and get tourist information.

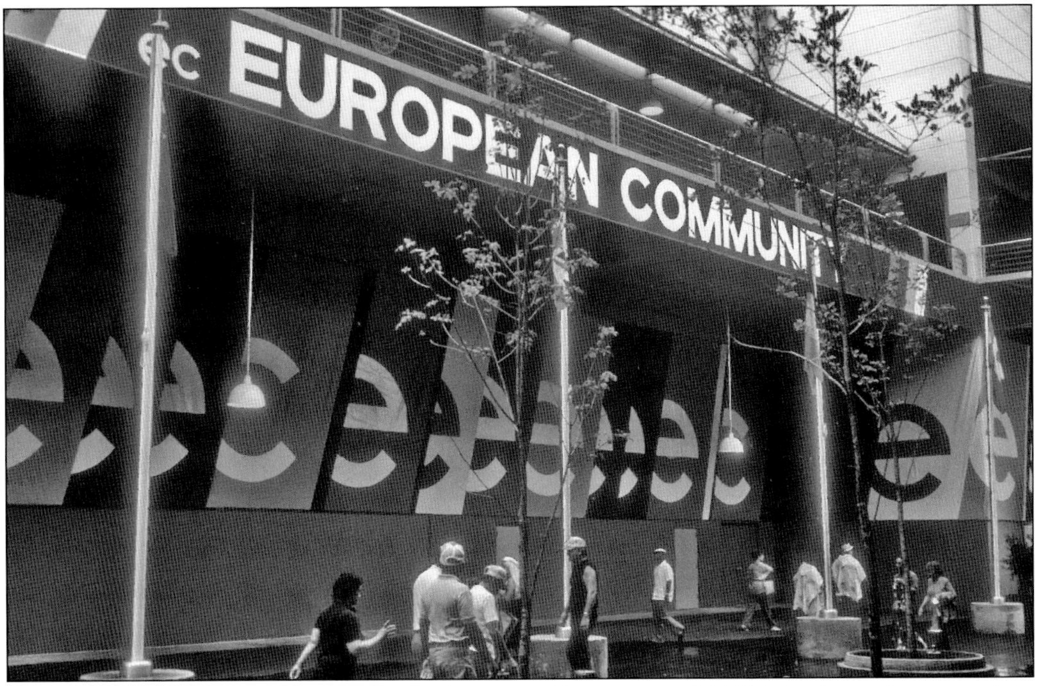

The 10 member nations of the European Economic Community (EEC) joined together to explain the successes they had achieved over the previous 25 years. In a not-so-subtle hint, it was suggested that the United States and the EEC nations could all benefit from closer ties made possible by tariff decreases. A 15-minute film showcased many of Europe's most important waterways and explained their value to commerce.

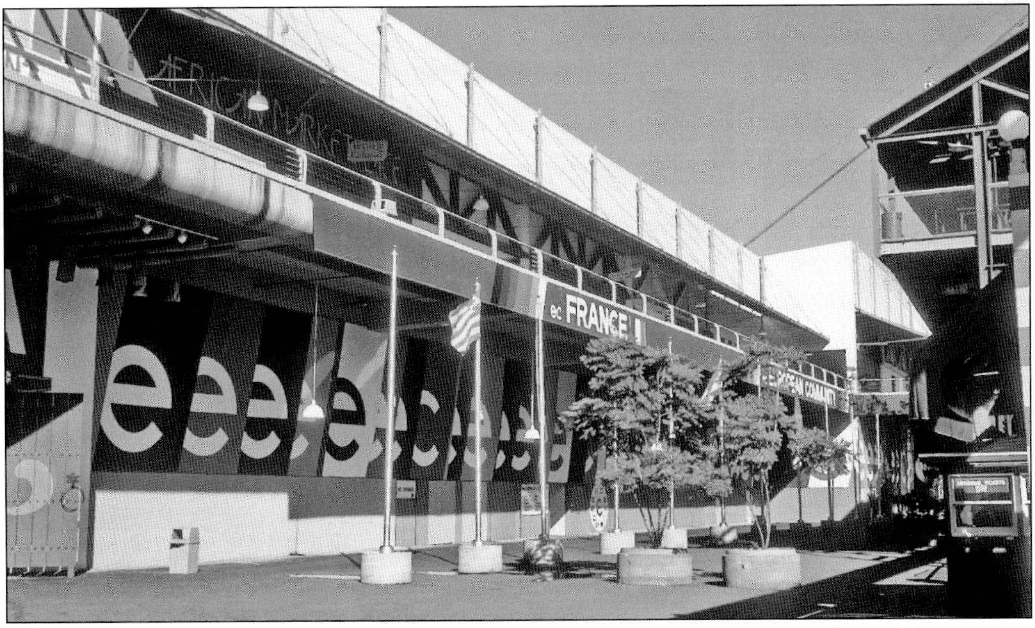

Although a member of the EEC, France had its own pavilion as well. Exhibits showed the progress French companies had made in improving water purification and in building hydroelectric systems and dams. French waterways, such as the Seine, were also seen in a more tourism-oriented display.

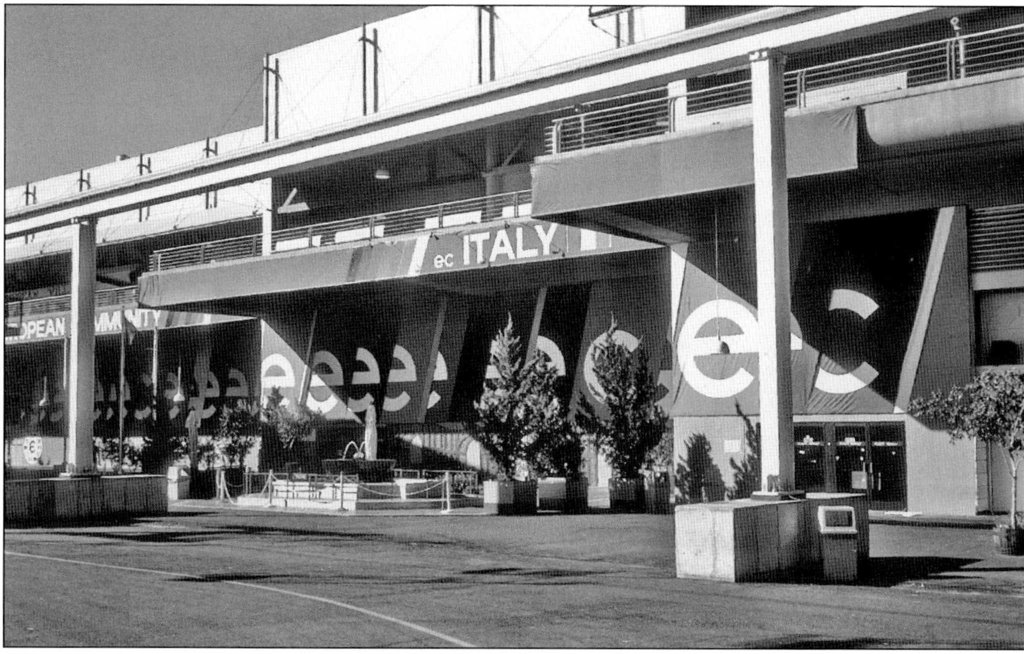

The other EEC member with its own pavilion was Italy. Less commercial than the main EEC displays or France, Italy provided a look at how early work to reclaim marshes and provide irrigation provided important lessons that are still valuable today. Many of these early constructions are now tourist attractions, such as the famous canal city of Venice.

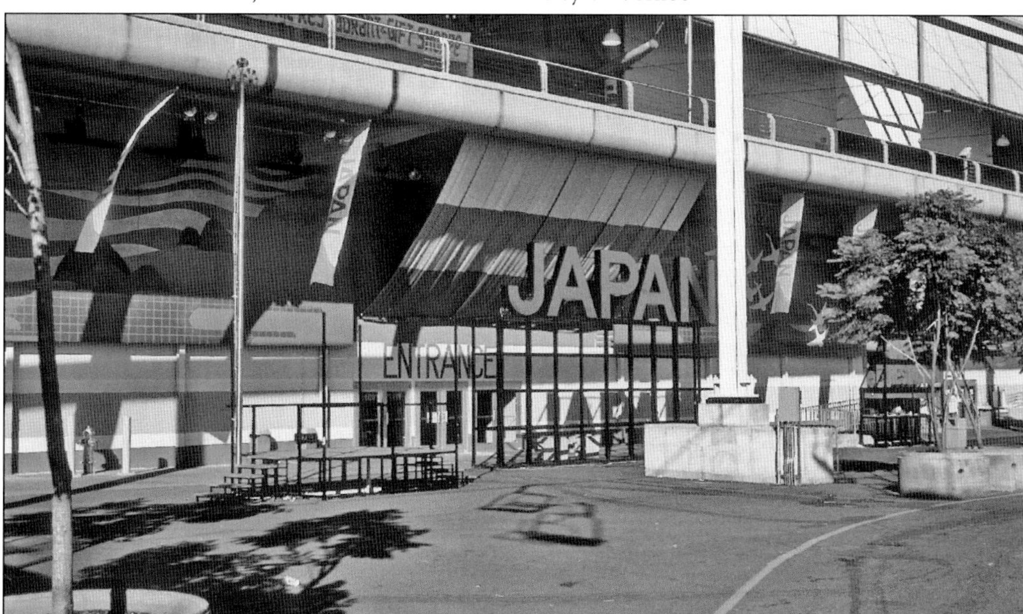

The Japan Pavilion included displays about how water has long been important to Japanese culture and industry. After a display on the use of water in Japanese gardens, a recreation of an early waterwheel designed with American assistance acknowledged the long relationship between the two countries. Other displays included Japanese computers and samples of many Japanese festivals.

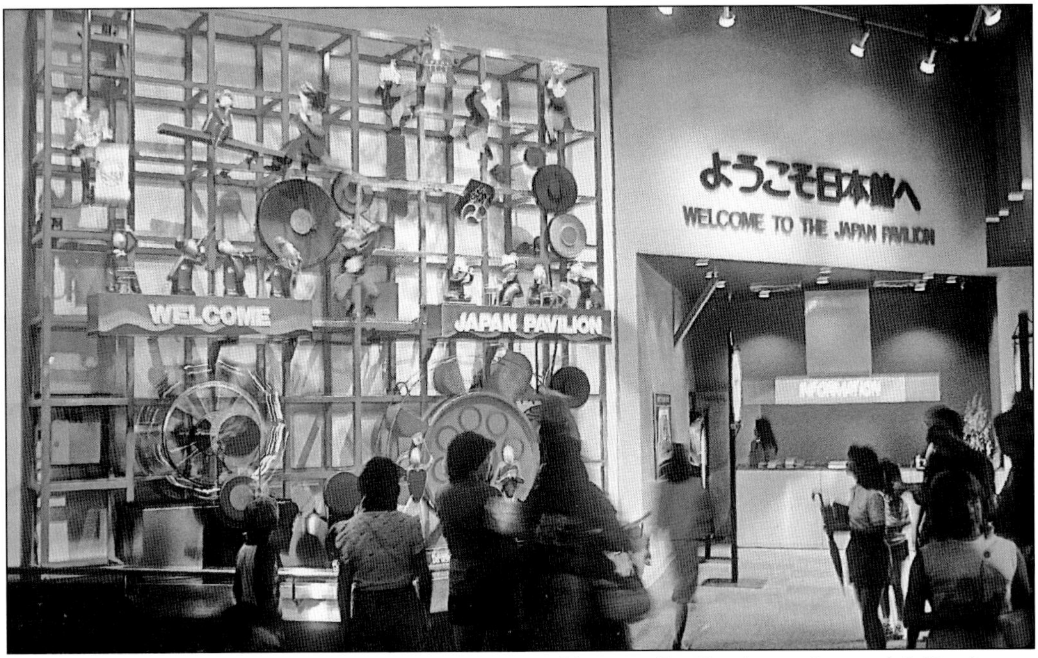

Guests entering the Japan Pavilion were entertained by a wall of lively mechanical dolls playing a variety of musical instruments. After traveling through several halls of films and displays about Japanese culture and business, the departing crowds were treated to a preview of Japan's Expo 85, the next world's fair.

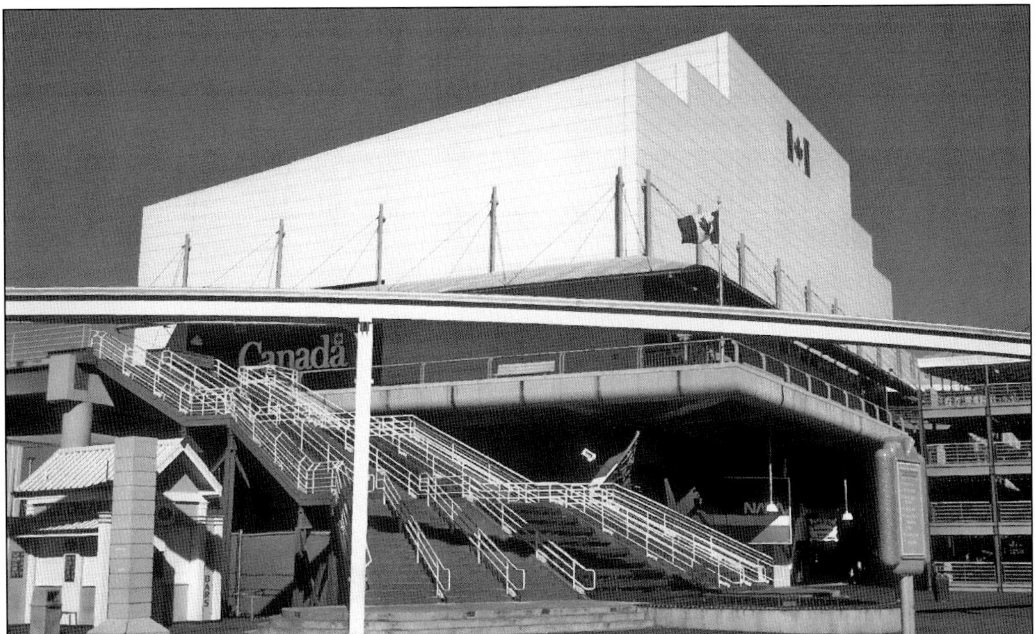

This large staircase at the southern end of the International Riverfront incorporated the fair's overall theme of water into its design with a long flowing fountain that cascaded down the middle into a small pool. Many weary visitors could be seen here resting their feet after a hot day at the fair.

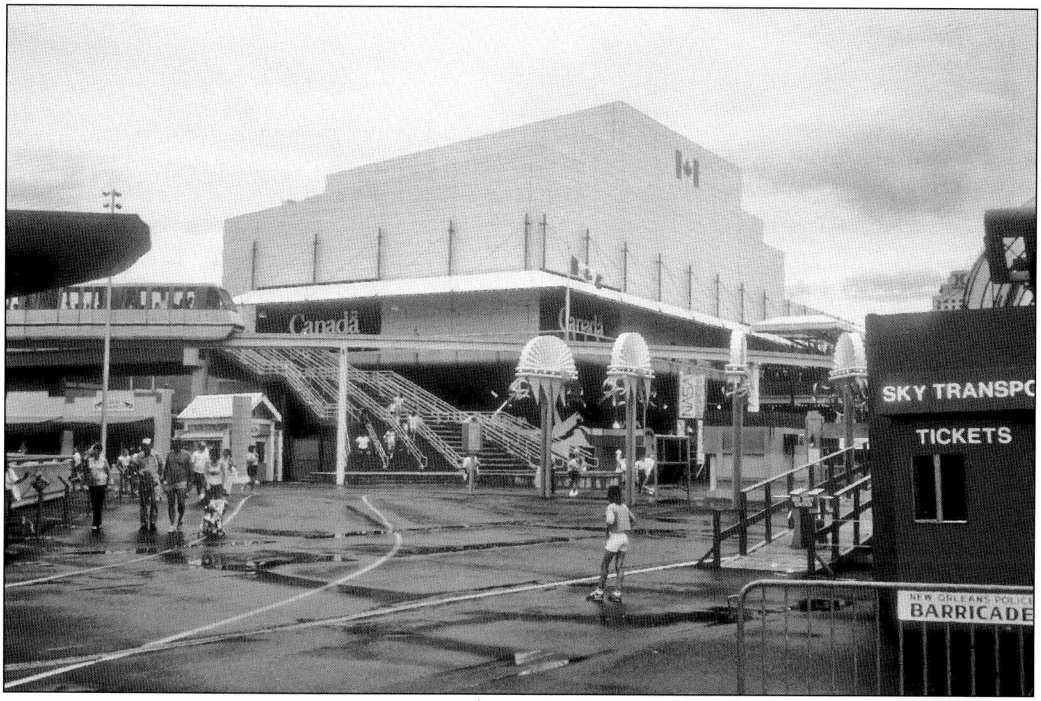

Canada had a major presence at the fair. Its 20,000-square-foot pavilion held four galleries that used a variety of audiovisual techniques. The highlight of the pavilion was a 15-minute IMAX film that explored how the country varied from region to region and changed between seasons. Similar to the Australia and Japan Pavilions, a presentation on Vancouver and Expo 86 was there to develop interest in that world's fair.

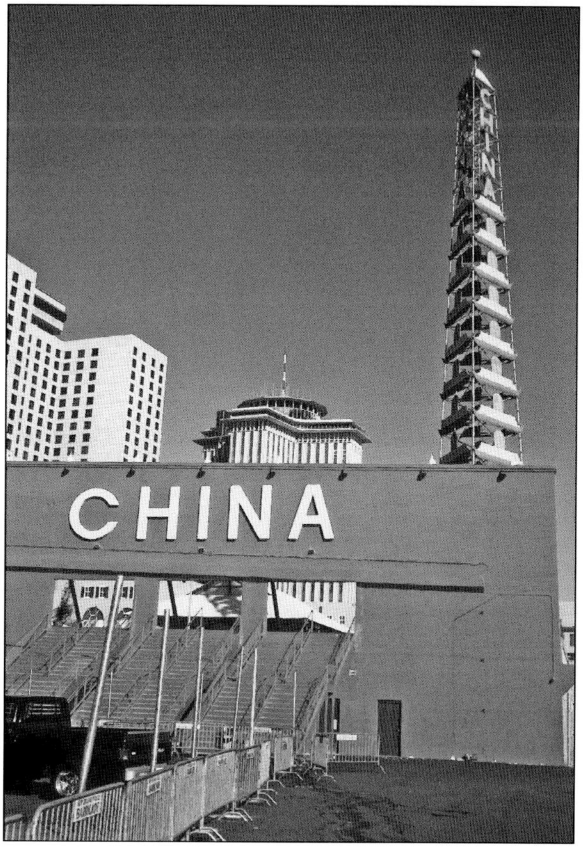

Located at the northern end of the International Riverfront, the China Pavilion was promoted as a display on the rich cultural history of the nation, with artists showcasing their talents. In reality, it was mostly a giant gift shop and is remembered by many as one of the biggest disappointments of the fair.

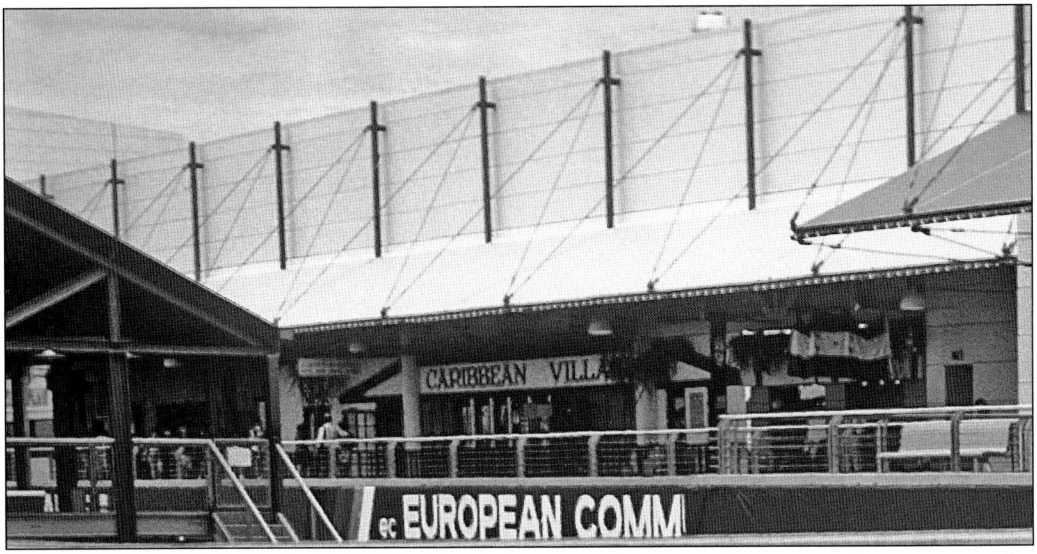

The Joint Caribbean Pavilion looked at the importance of water to the nations that comprise the Caribbean Basin. Water is an important part of life for those living on the island nations, and it also plays a critical role for those in Central America, as seen in a recreated rain forest.

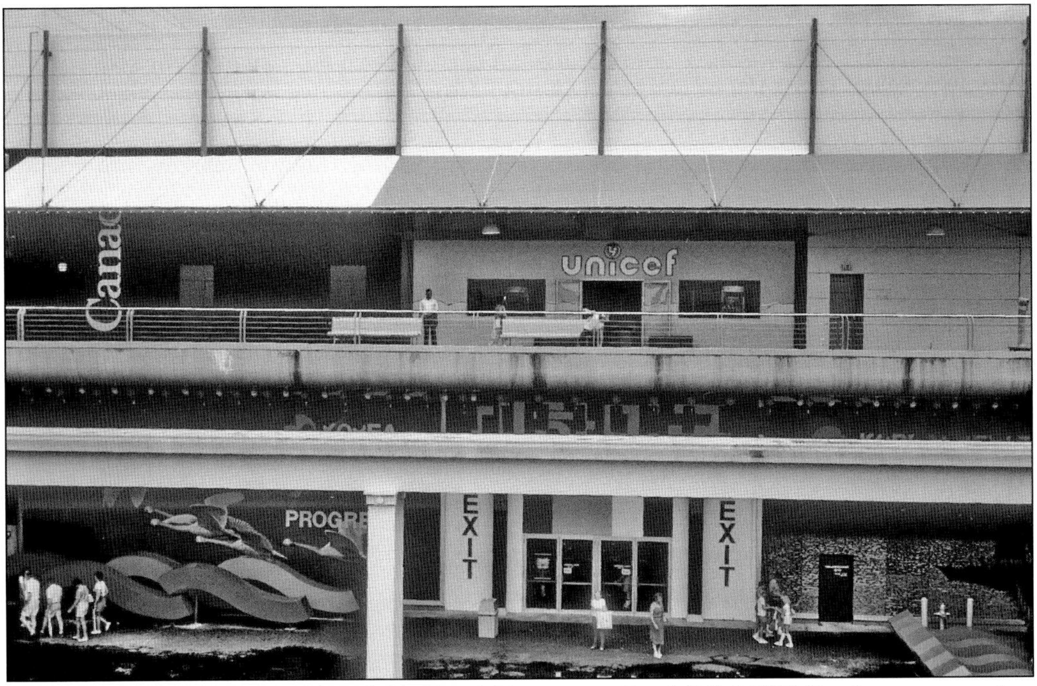

The United Nations International Children's Emergency Fund (UNICEF) Pavilion was another late addition to the fair. While much of the fair was geared towards adults, the UNICEF Pavilion was kid-friendly, with storytelling sessions, songs, and artwork by children from around the world. A presentation on the worldwide charitable organization's programs encouraged visitors to make donations at the fair or at home.

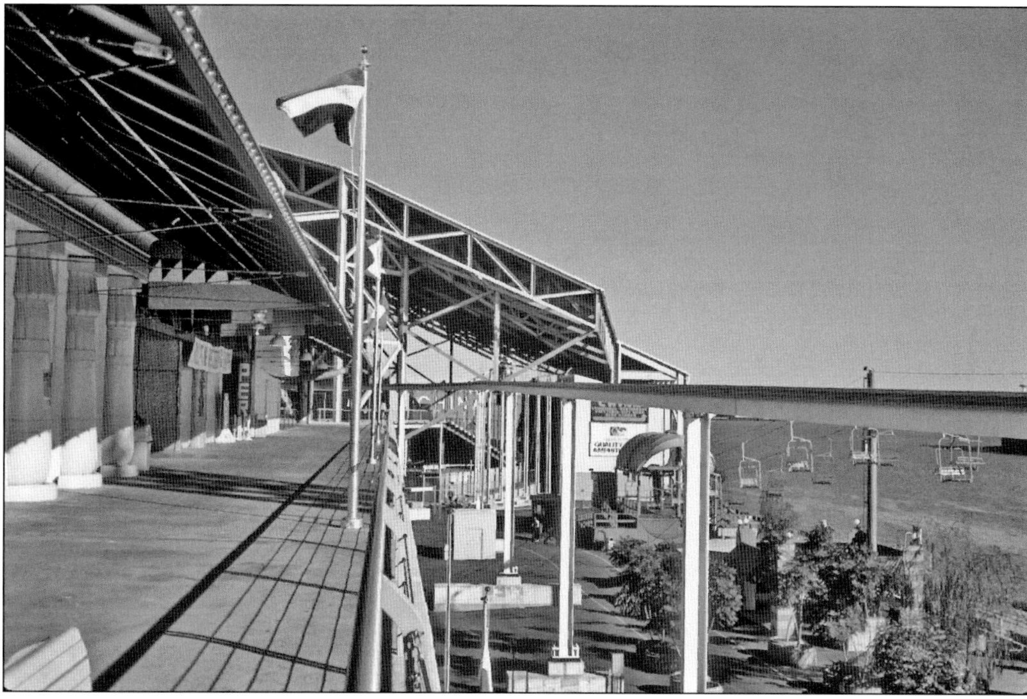

Peru was another of the international exhibitors, and its displays focused on the importance of water in the coastal, Andes, and Amazon Basin regions. Films and displays showed how water helped shape the early Peruvian cultures and remains an important element today. The Amazon Basin, for example, is a critical part of the Earth's ecosystem and has an impact on the whole world.

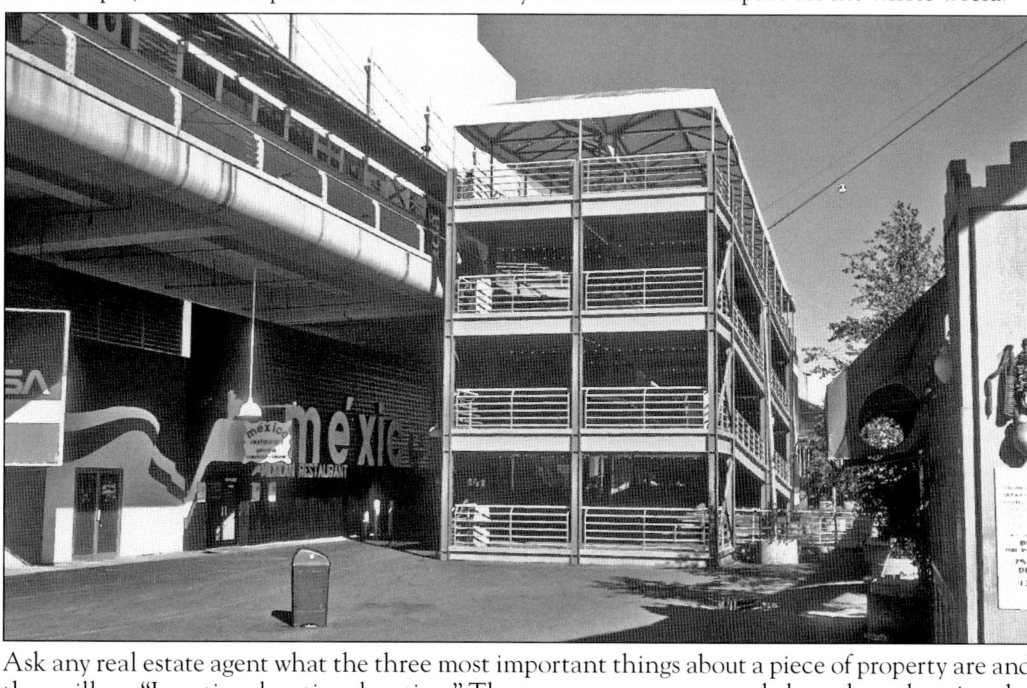

Ask any real estate agent what the three most important things about a piece of property are and they will say "Location, location, location." That message was apparently lost when planning the Mexico Pavilion, for it was all but buried behind the ramps of the monorail station.

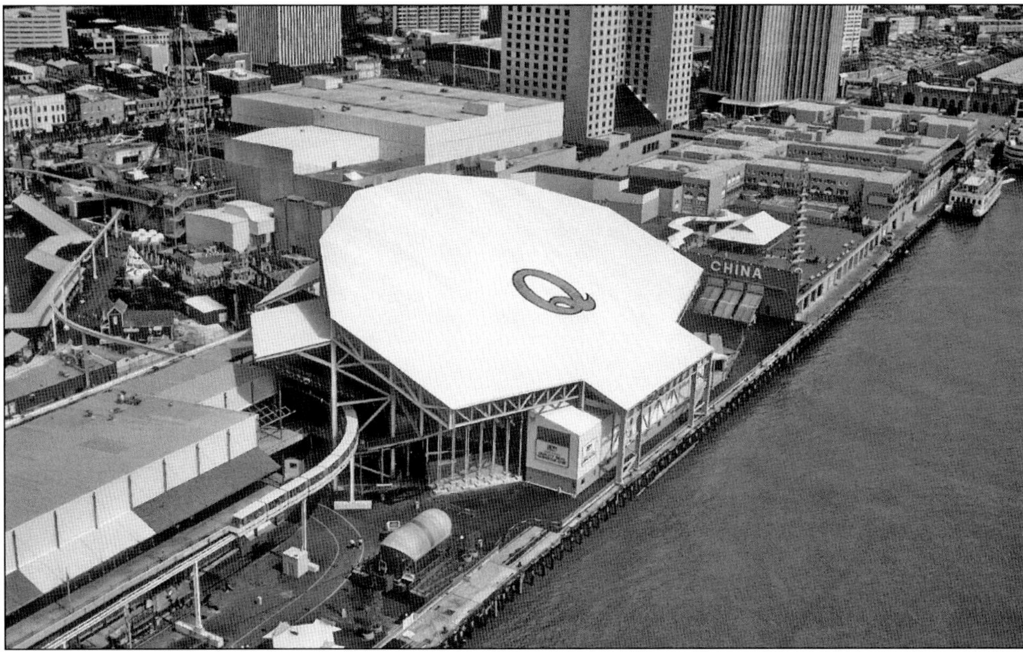

One special aspect of the fair was the live performances held at various venues. The largest was the Liggett and Myers Quality Seal Amphitheatre, which seated 5,500. Designed by famous architect Frank O. Gehry, the amphitheater was especially popular in the evening, when entertainers would often perform to sold-out crowds.

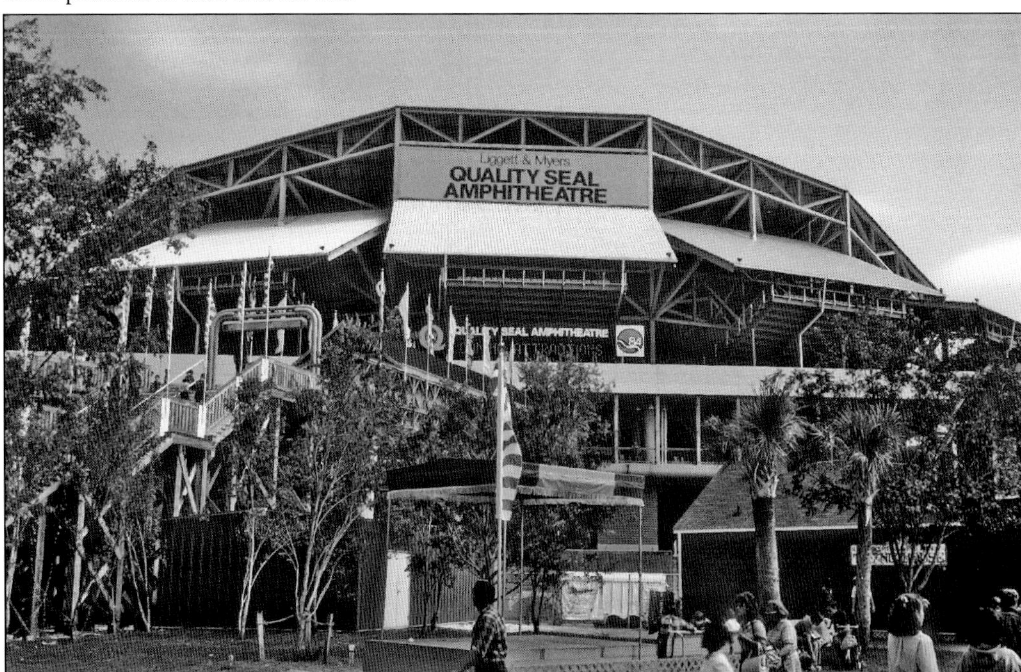

Seen here from Centennial Plaza, the amphitheater was a relatively late addition to the fair that did not appear on many of the original site plans or concept art. While a more ambitious roof design had to be scrapped, the final product became one of the busiest sites at the fair.

The plans for the International Riverfront included a steady parade of ships of all types from around the world; free tours were provided as part of the fair. Operation Ship, as the plan was called, never really succeeded, because few foreign ships participated. Most of the ships that did make it were U.S. Navy or Coast Guard vessels.

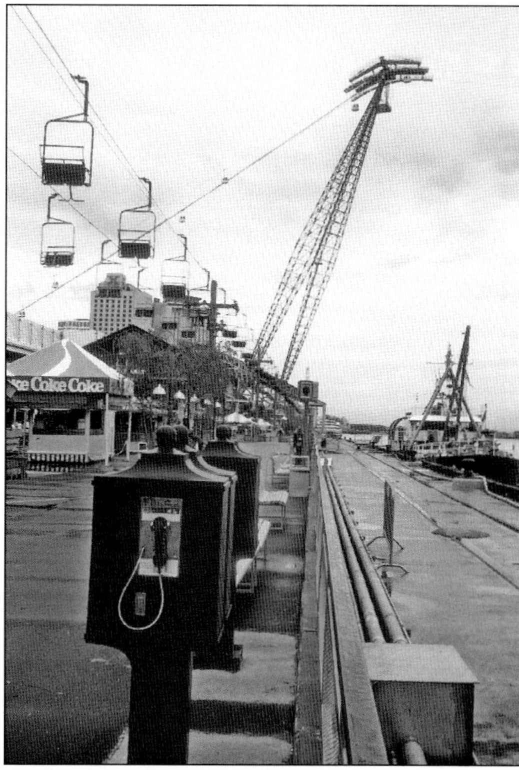

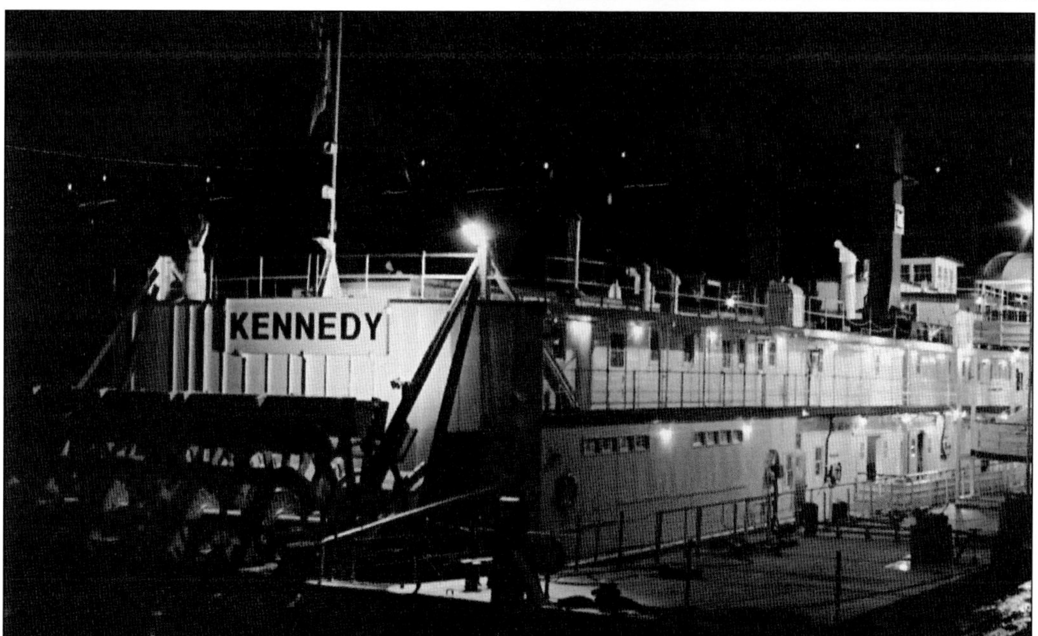

One vessel was there for the duration of the fair. The U.S. Army Corps of Engineers brought the restored dredge *Kennedy*, launched in 1932, to demonstrate its role in keeping the Mississippi River and other waterways open for shipping. The displays also included a history of the corps and how it helped to reduce damage from flooding on the Mississippi. (Courtesy of James Pilet.)

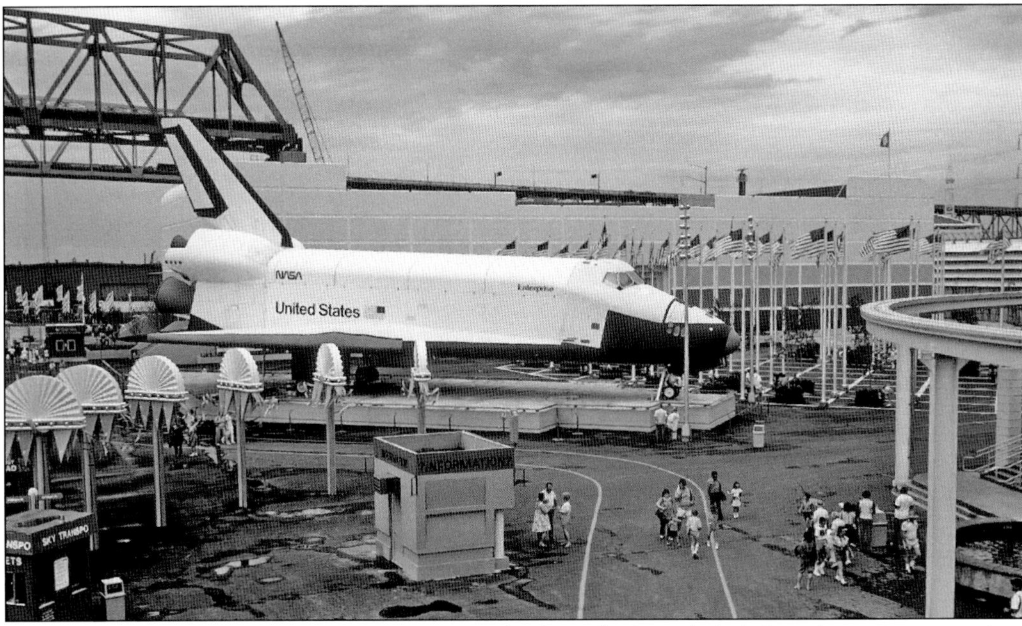

One of the high points of the fair was NASA's display, which included the orbiter *Enterprise* from the Space Shuttle fleet. NASA had originally considered taking the *Enterprise* to the 1982 World's Fair in Knoxville, but a series of logistical obstacles blocked that plan. Happily for the organizers of this fair, the location was ideal for transporting the *Enterprise* by barge. The *Enterprise* is seen here in front of the United States Pavilion.

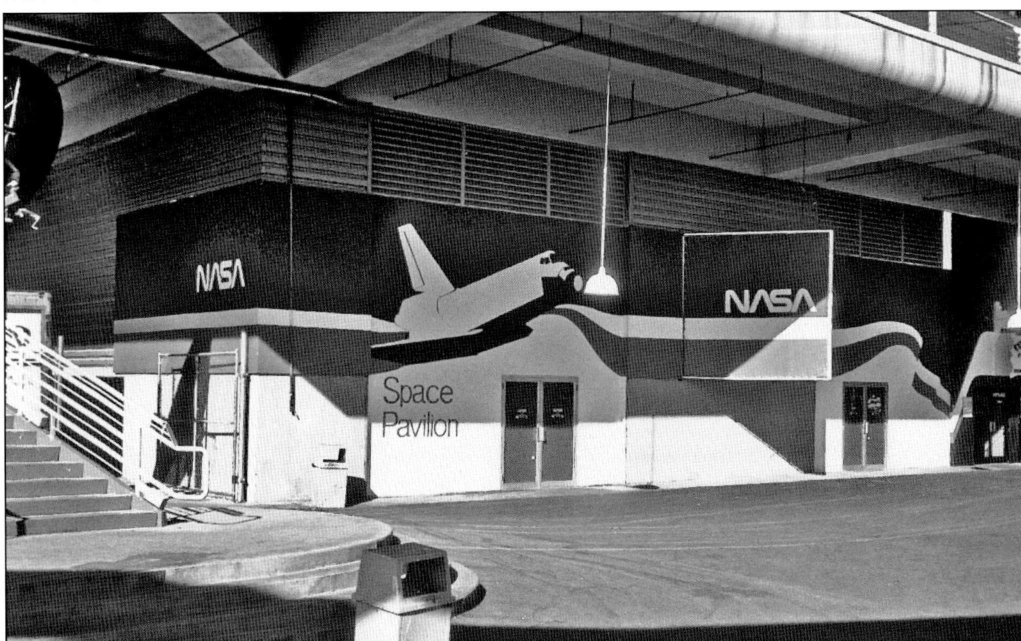

Although most people would call the *Enterprise* a space shuttle, the orbiter is only part of the launch vehicle that makes up the actual shuttle. Inside the Space Pavilion, displays on the shuttle program explained how all its parts were necessary to get the orbiter into space. Exhibits also explained the importance of weather satellites and other NASA projects.

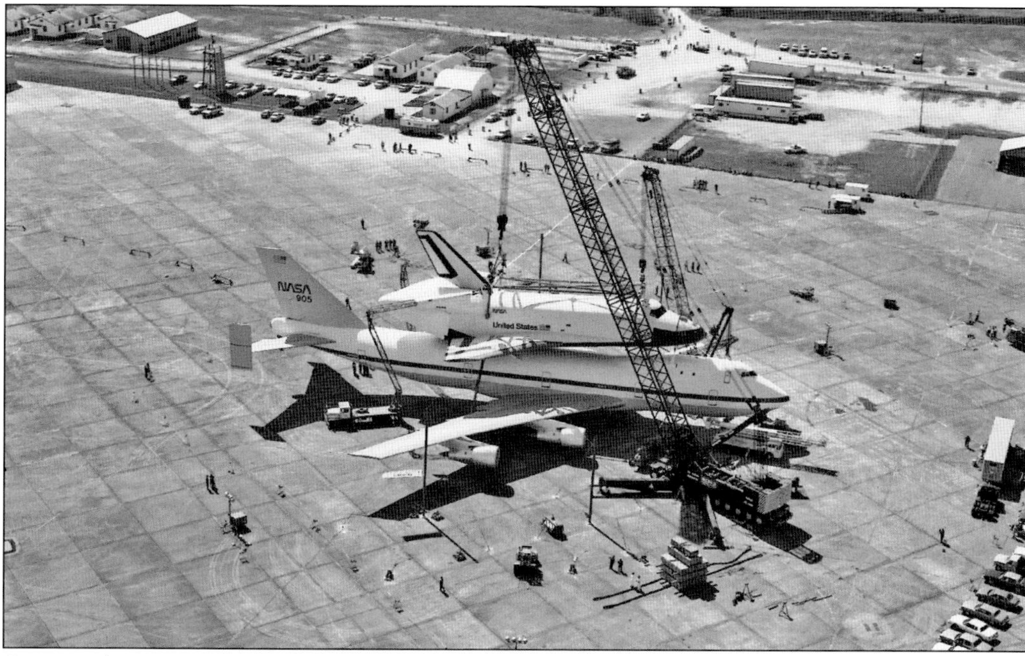

Getting the *Enterprise* to the fair site was quite a challenge. There were no roads between the New Orleans airport and the fair that could be used to transport the orbiter, so NASA decided to get it there by river. The first stop was Mobile, Alabama, where cranes were used to lift the *Enterprise* from its Boeing 747 carrier. (Courtesy of NASA.)

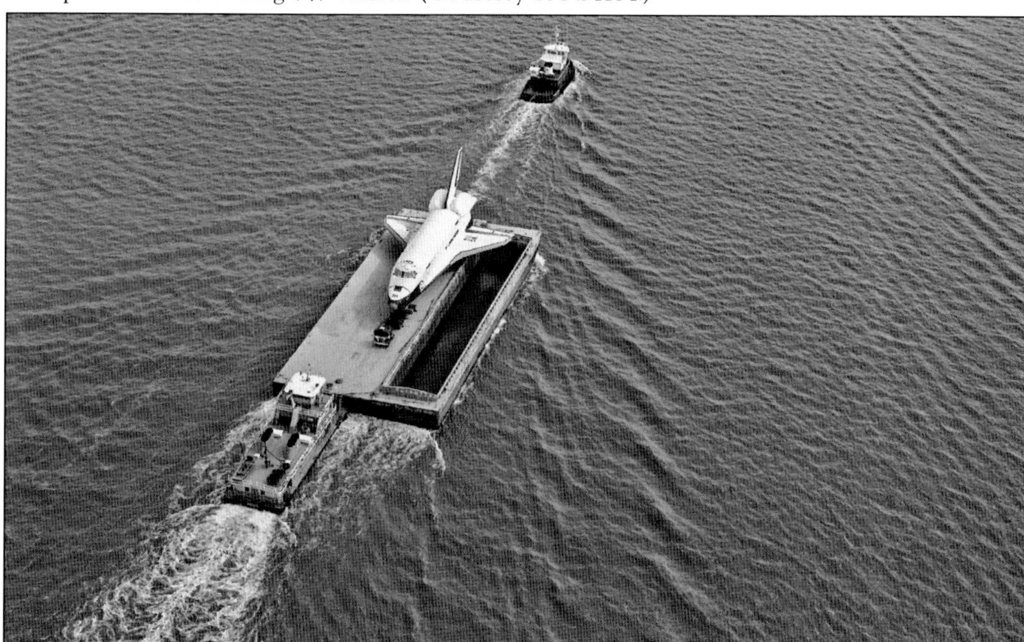

The barge was then towed through the Gulf of Mexico to the Mississippi River and from there to the fair site. The weather had to be carefully monitored, for NASA did not want to take any chances with its prized cargo. The *Enterprise* and barge were so large that the tractor-trailer sharing the barge can barely be seen. (Courtesy of NASA.)

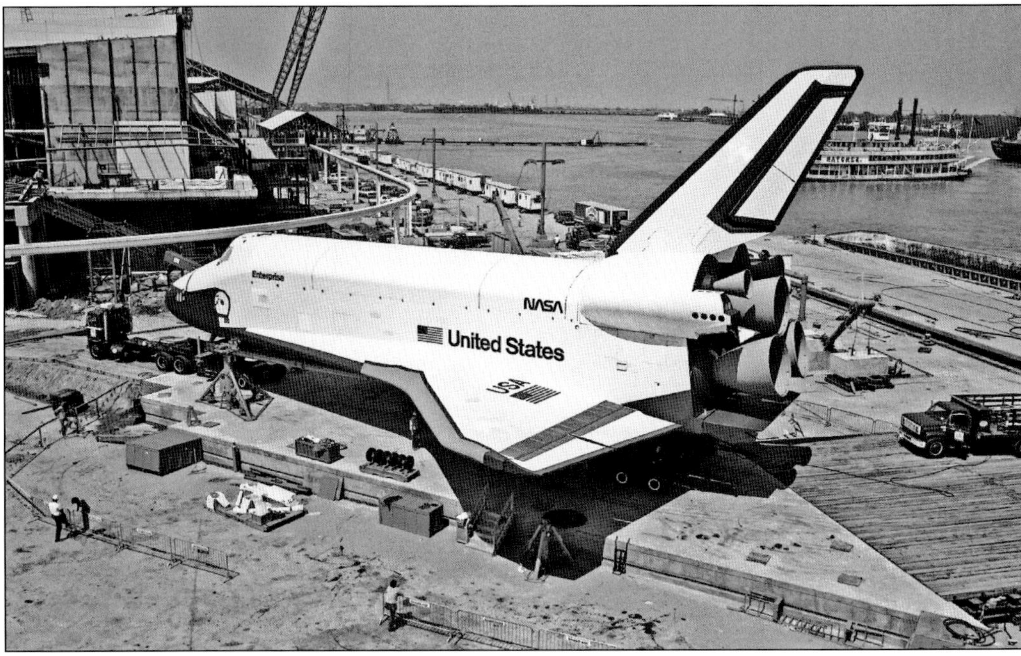

A temporary roadway is used to get the *Enterprise* off the barge and onto its concrete display stand as work continues on the rest of the International Riverfront. It took months of planning for the 11-day trip, but appreciative crowds lined up to see the spaceship at the fair. (Courtesy of NASA.)

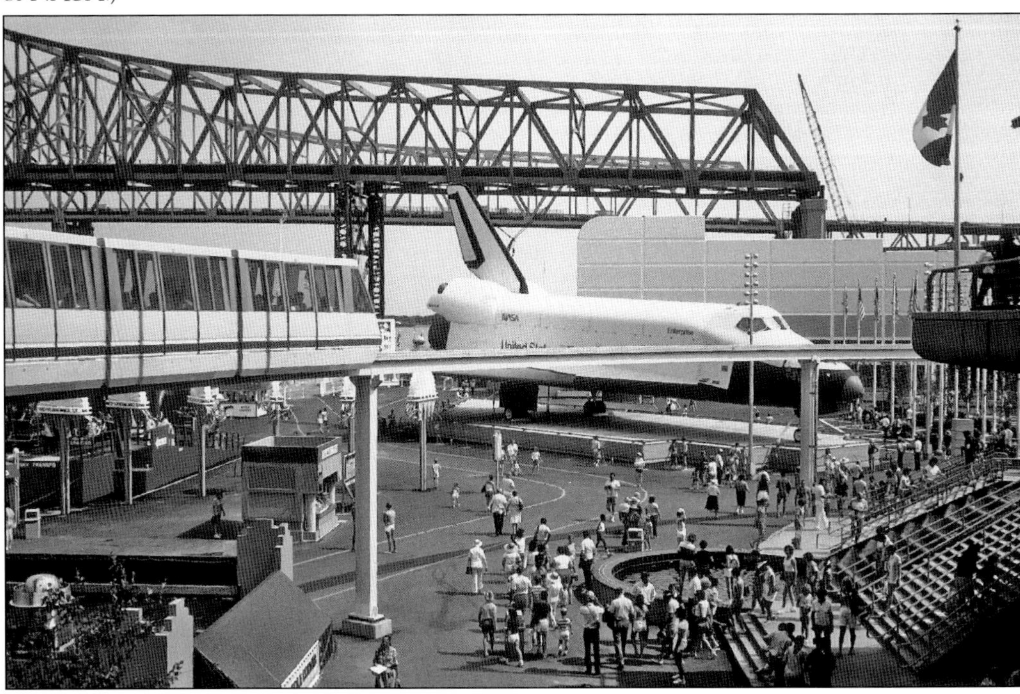

Two advanced forms of transportation can be seen from the upper level of the International Riverfront as the monorail passes the *Enterprise*. A more traditional method, the Greater New Orleans Bridge, is seen under construction in the rear.

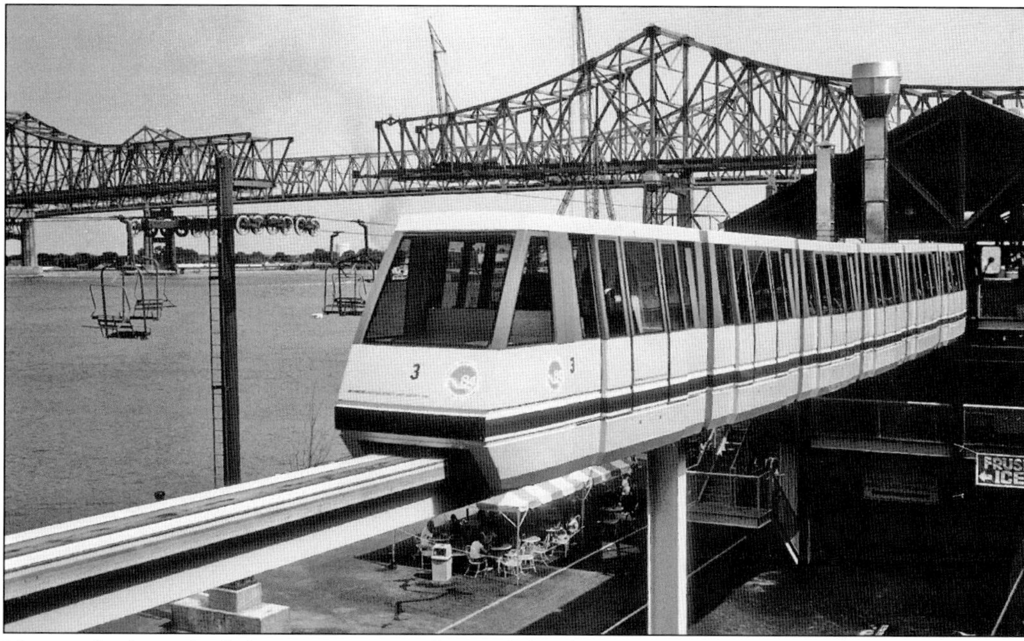

Monorails always look futuristic, even today, and this fair's was a popular subject for photographers. Traveling at a top speed of 10 miles per hour, the monorail could complete a trip around the fair in approximately 12 minutes. The train here is just outside the International Riverfront station.

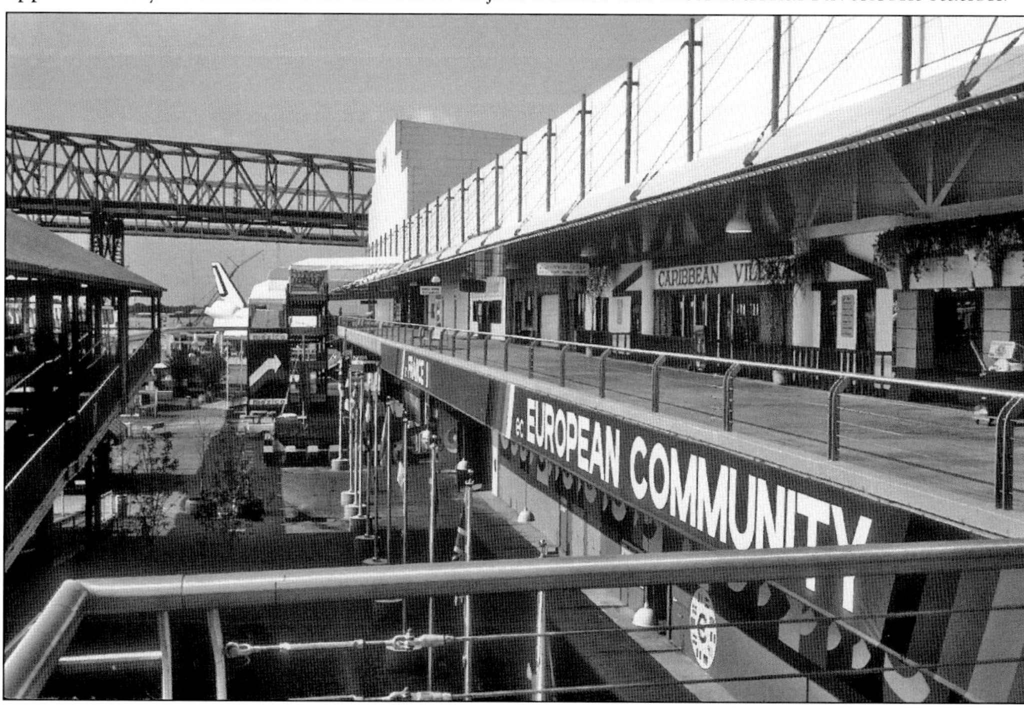

Despite the various transit systems, most guests toured the area on foot, so wide walkways were added to the upper level to accommodate the crowds predicted by the fair organizers. While attendance was lower than expected, this view of a deserted area was actually taken before the gates opened for the day.

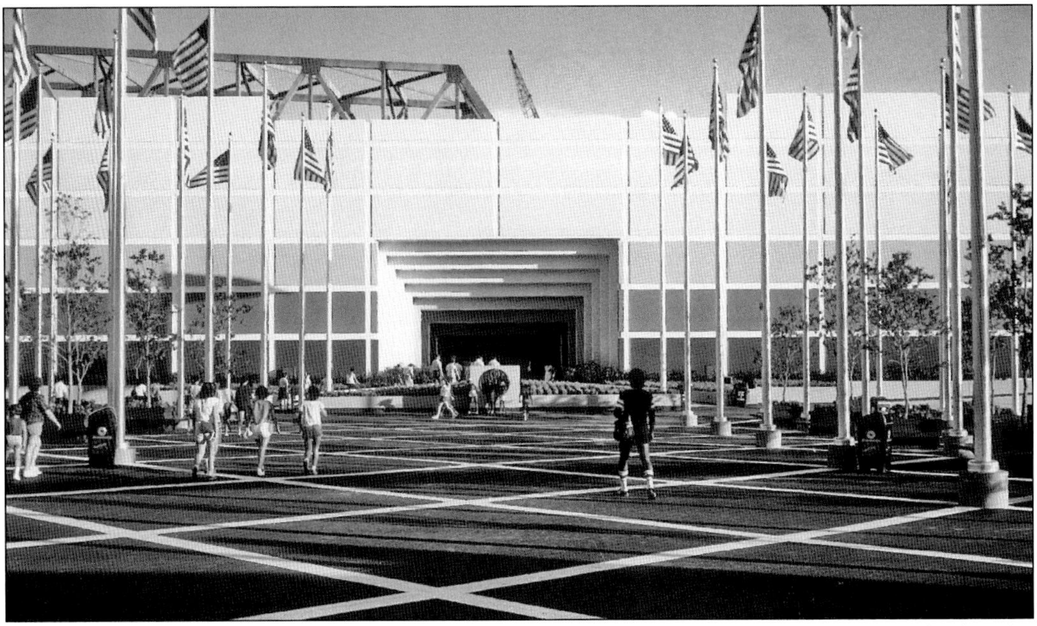

The United States Pavilion was a striking sight, with dozens of American flags lining both sides of the entrance. Inside guests were entertained by a 20-minute 3-D film titled *Water, the Source of Life*, which included views of some of the country's most famous rivers and waterfalls. High-tech displays explained the importance of water in subjects as varied as the human body and power plants.

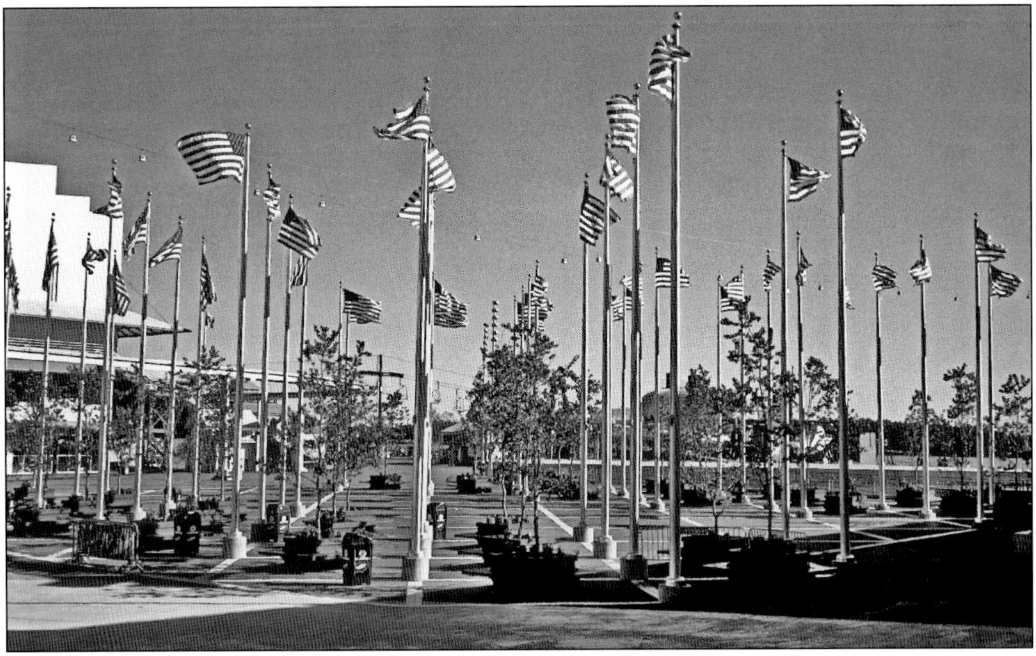

The planters at the base of the flagpoles outside the United States Pavilion were the only sign of greenery in the International Riverfront area. They turned what could have been a rather uninteresting area into a colorful display that helped draw the eye, and visitors, towards the pavilion entrance.

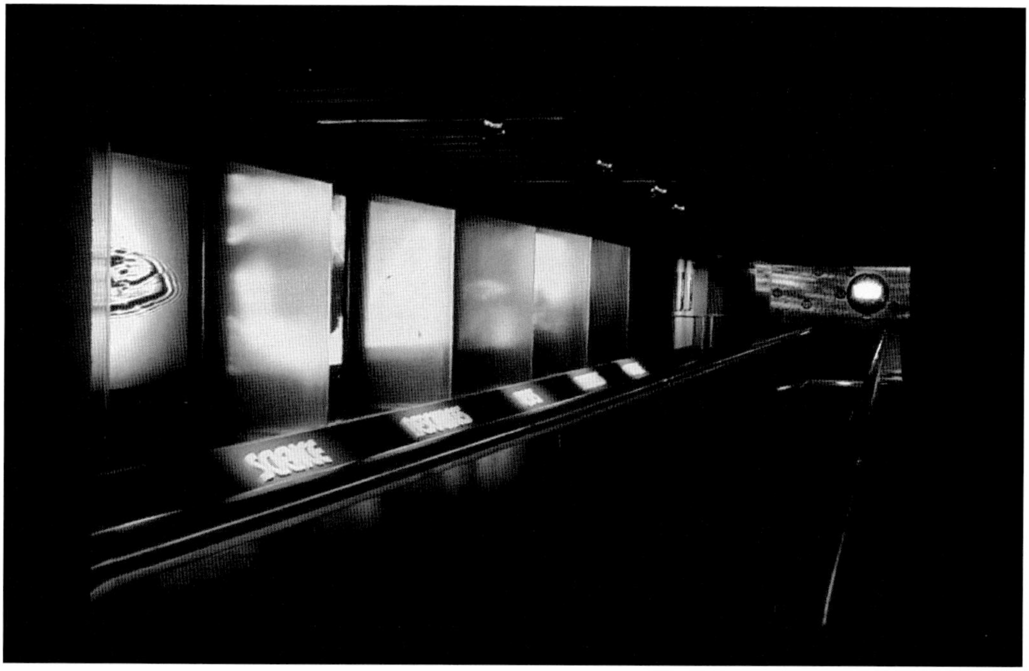

Once inside the United States Pavilion, guests stepped onto a moving sidewalk that carried them past displays depicting the many reasons water is important to us all. Curtains of falling water separated the photographs on view in each section. (Courtesy of Barry Howard.)

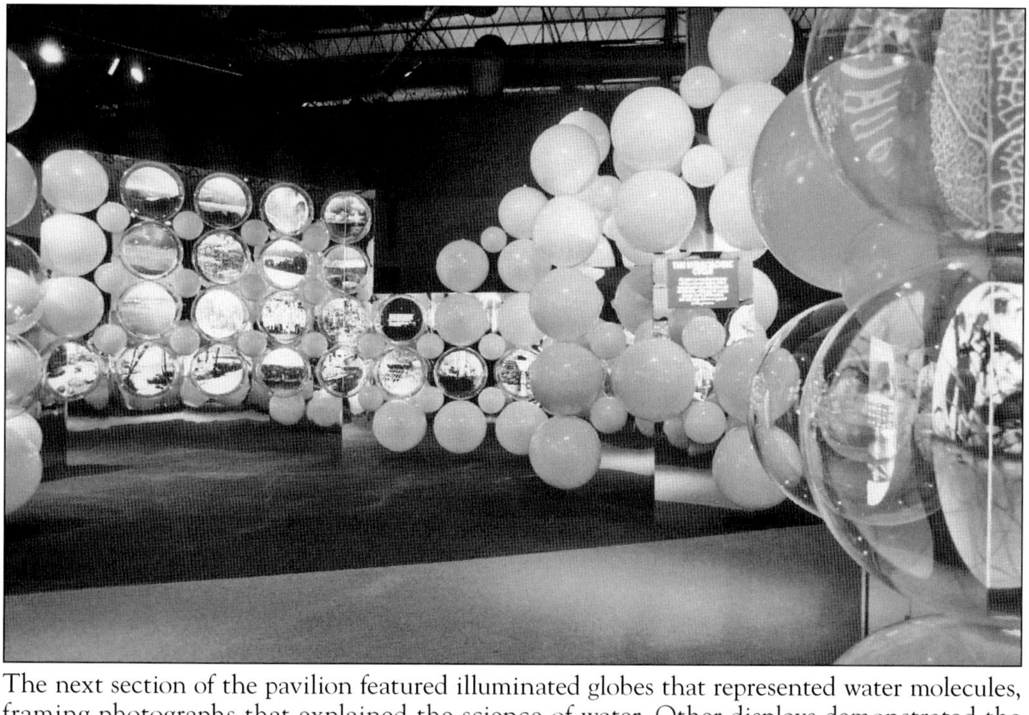

The next section of the pavilion featured illuminated globes that represented water molecules, framing photographs that explained the science of water. Other displays demonstrated the amount of water consumed in everyday life, in manufacturing, and in agriculture. (Courtesy of Barry Howard.)

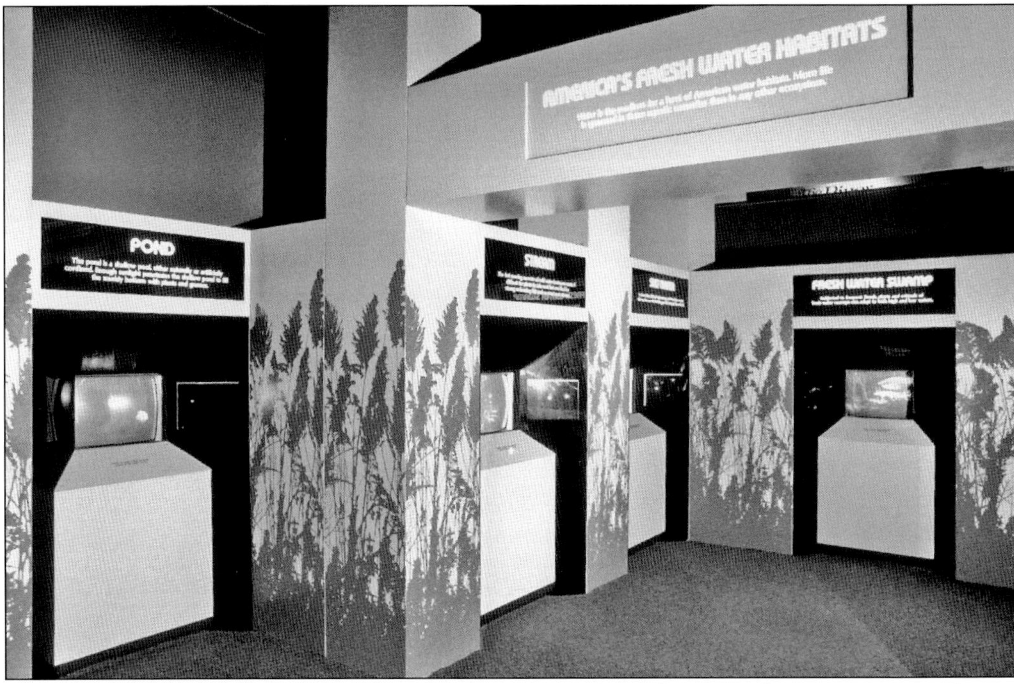

Video displays and interactive games educated visitors about the many forms water can take and its importance to daily life. This section explained the differences between various bodies of water such as ponds, streams, saltwater marshes, and freshwater swamps. (Courtesy of Barry Howard.)

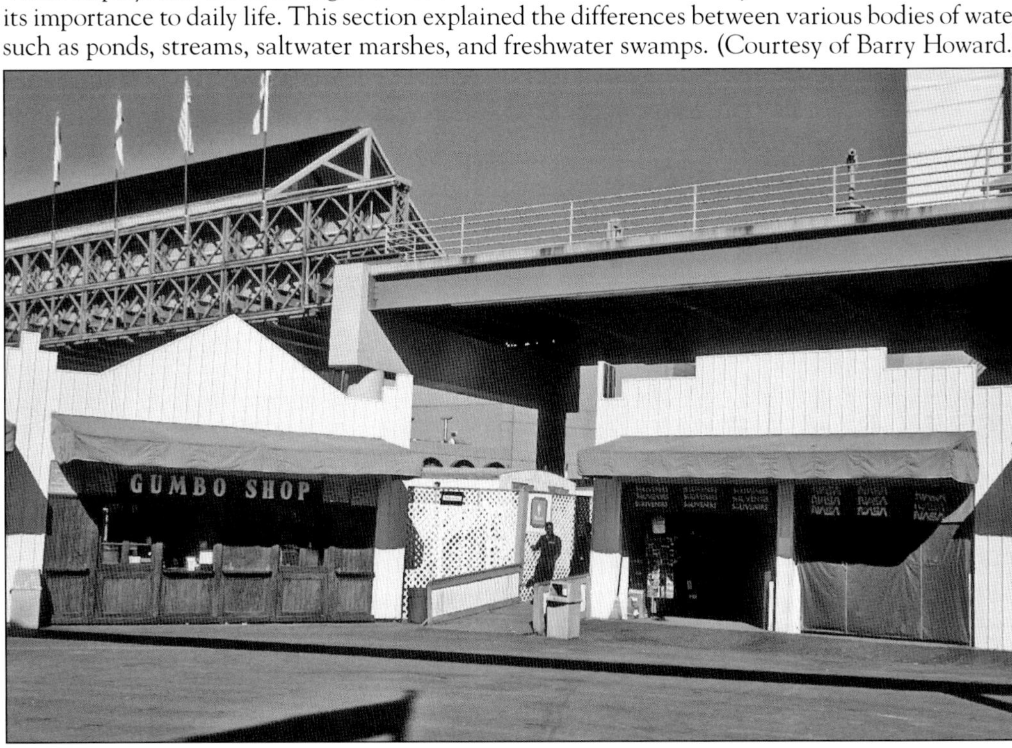

Part of the fun at world's fairs is trying foods from around the world. For New Orleans, it was only natural that in addition to the usual American fare of hamburgers and hot dogs, visitors could also enjoy gumbo, crawfish, alligator, and other local treats.

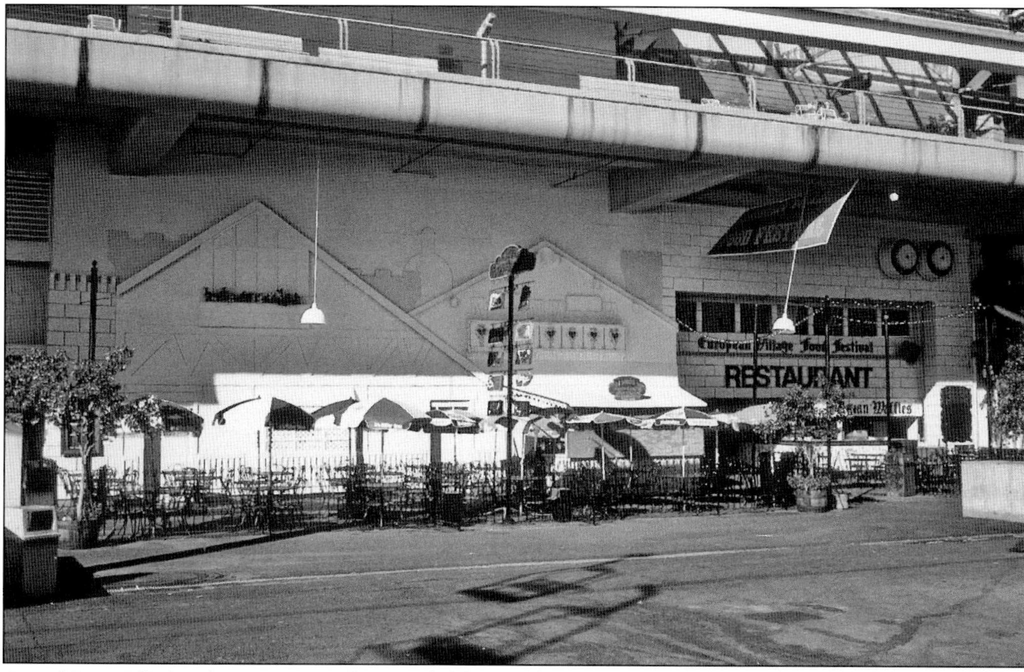

An old world's fair favorite was also available. With Belgian waffles having been such a hit at the 1964–1965 New York World's Fair, it was inevitable that they would also be sold at this fair. The European Village Food Festival was just one of the locations offering this tasty treat.

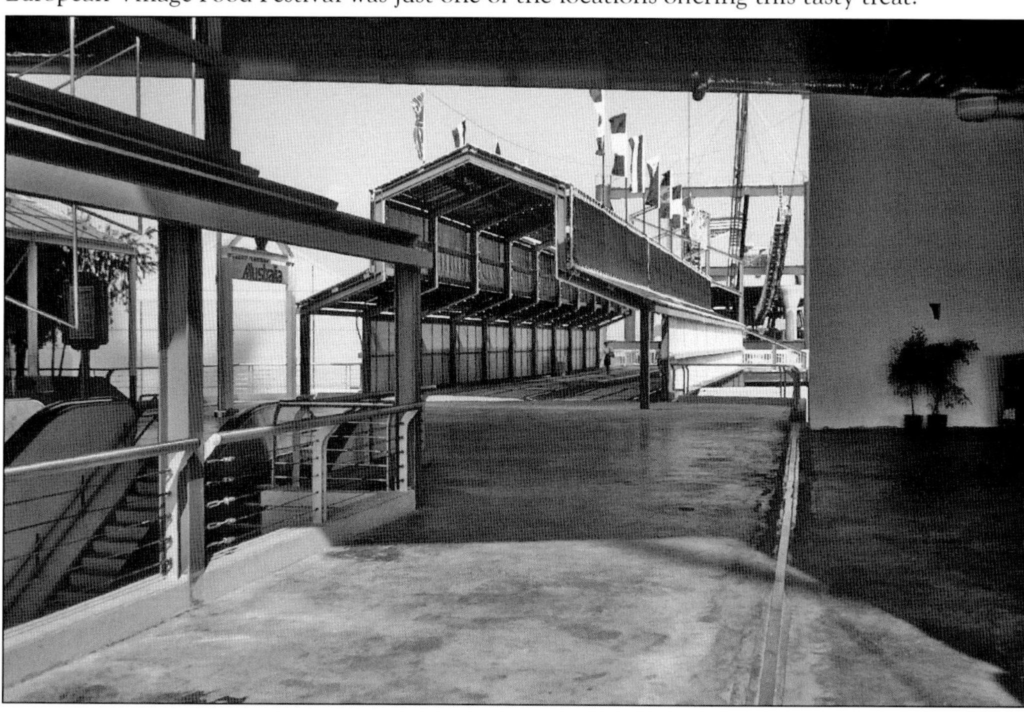

Two bridges linked the International Riverfront to the rest of the fair. This one, which crossed over to Centennial Plaza, was situated at what would have been the entrance to the Tower of New Orleans. The failure of that project left only a barren concrete floor filling the area.

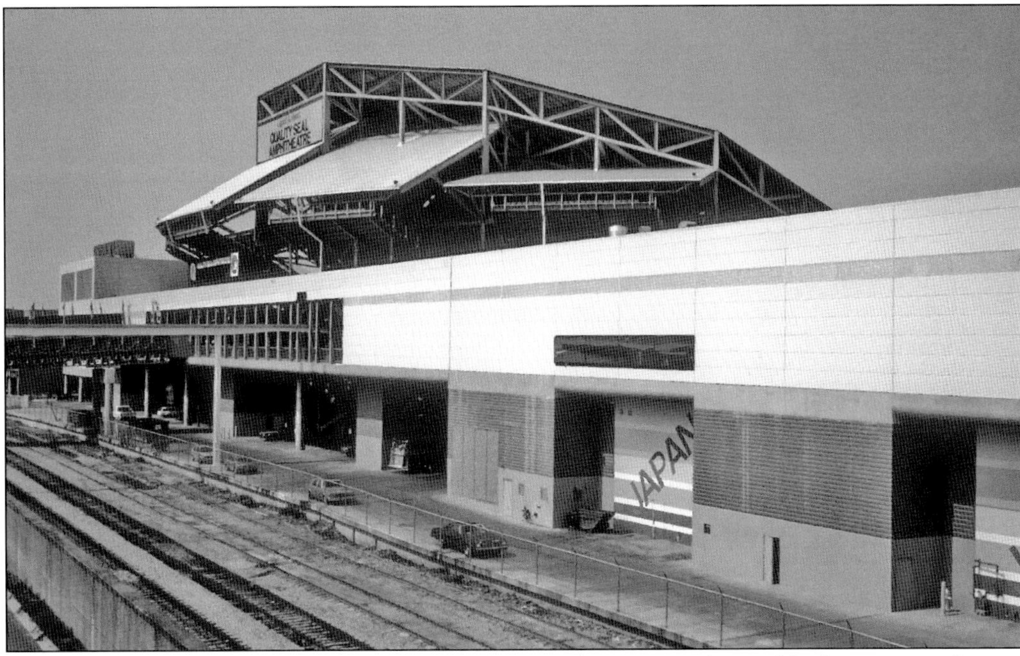

The rear of the International Riverfront clearly shows how it was planned for later reuse of a more industrial nature. Visitors crossing the bridges over the railroad tracks were treated to a view that looked more like a warehouse than a world's fair. While the Wonderwall was used to camouflage sights like this elsewhere in the fair, no such effort was made here.

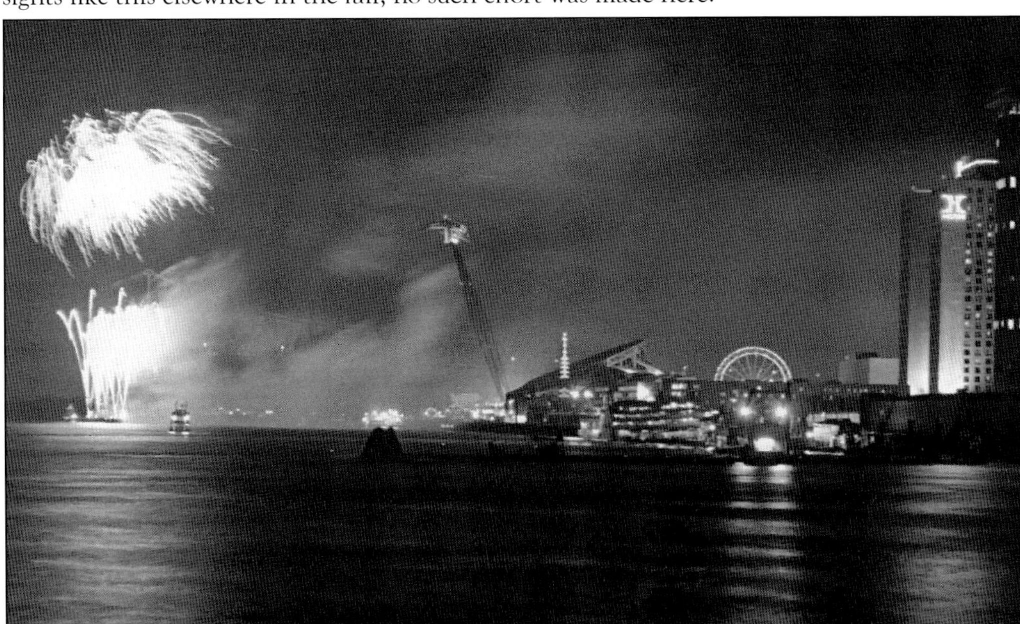

One of the highlights of the fair was the nightly fireworks show. Launched from a barge in the river, the fireworks are still fondly recalled by many visitors to the fair. The festive lighting on the pavilions and Giant Wheel all contributed to a look that was quite different from the pre-fair days. (Courtesy of James Pilet.)

Eight

THE WONDERWALL

The Wonderwall started out as a solution to an architectural problem: How to dress up a street dividing the ultramodern Exhibition Center from a row of century-old renovated warehouses.

—Official Guidebook

One can only wonder how the fair planners felt when they first studied the site, for it was full of obstacles and challenges. An active rail line ran through it, meaning that a method was needed to get guests safely across it without interrupting the trains. The fair would be a mix of old buildings and new, making it a challenge to bring the two together. Possibly worst of all, though, was the visual blight caused by a number of high-tension electrical cables that ran through the middle of the site. A solution was needed to reduce the negative impact these intrusions would have brought to the fair experience.

The answer was the Wonderwall, a half-mile-long collection of shapes and colors that shouldn't have worked together but did. The project was led by Charles Moore and William Trumbull, who also designed the fair's Centennial Pavilion. Moore had previously designed the Piazza d'Italia, an urban square in New Orleans that mixed architectural and cultural elements from across Italy. An unusual collection of shapes, colors, and materials, the piazza won critical acclaim for Moore's approach to rejuvenating a barren piece of real estate.

The Wonderwall built upon the piazza experience, with Moore crediting the works of 18th-century Italian artist and architect Giovanni Battista Piranesi as his inspiration for the new venture. Additional artists were invited to contribute to the design, and the result was a massive piece of art that was different everywhere one looked. In some places, the Wonderwall was wide enough to house shops and even a radio station. In others, it was more of a decorative screen. The height varied, the colors varied, and even the basic shapes varied, leading to one critic's description of it as an "architectural gumbo." The *New York Times* described it as "a visual roller coaster in which giant alligators and pelicans leaped over classical temples toward a giant Ferris wheel."

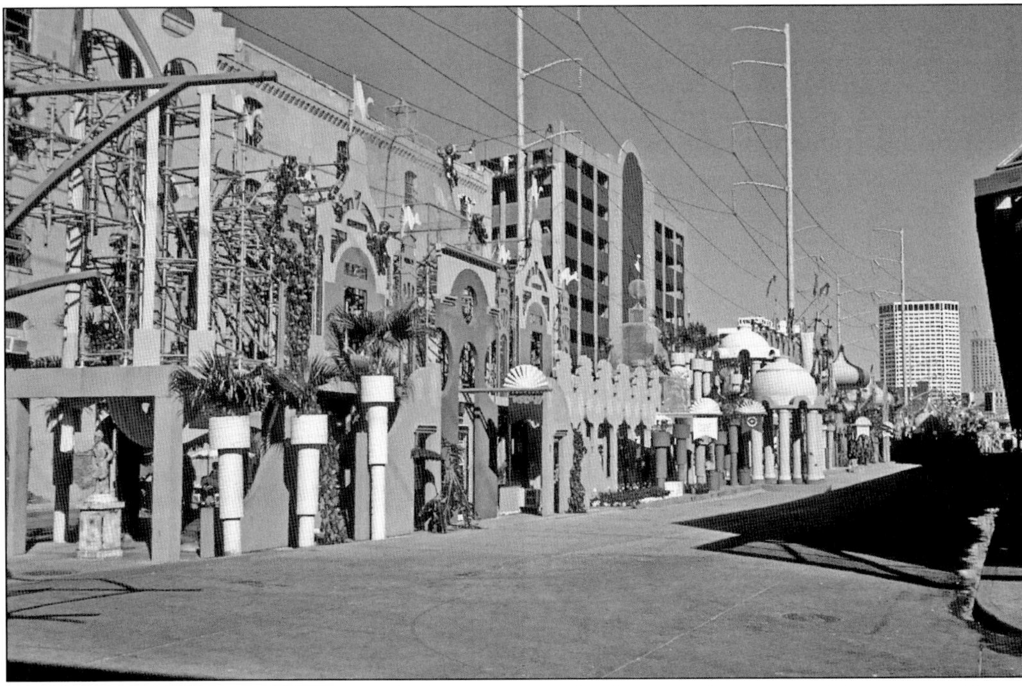

Stretching down the middle of South Front Street, which is now known as Convention Center Boulevard, this section of the Wonderwall does not look very substantial, but it was actually large enough to house a radio station, which was located in the center of the photograph.

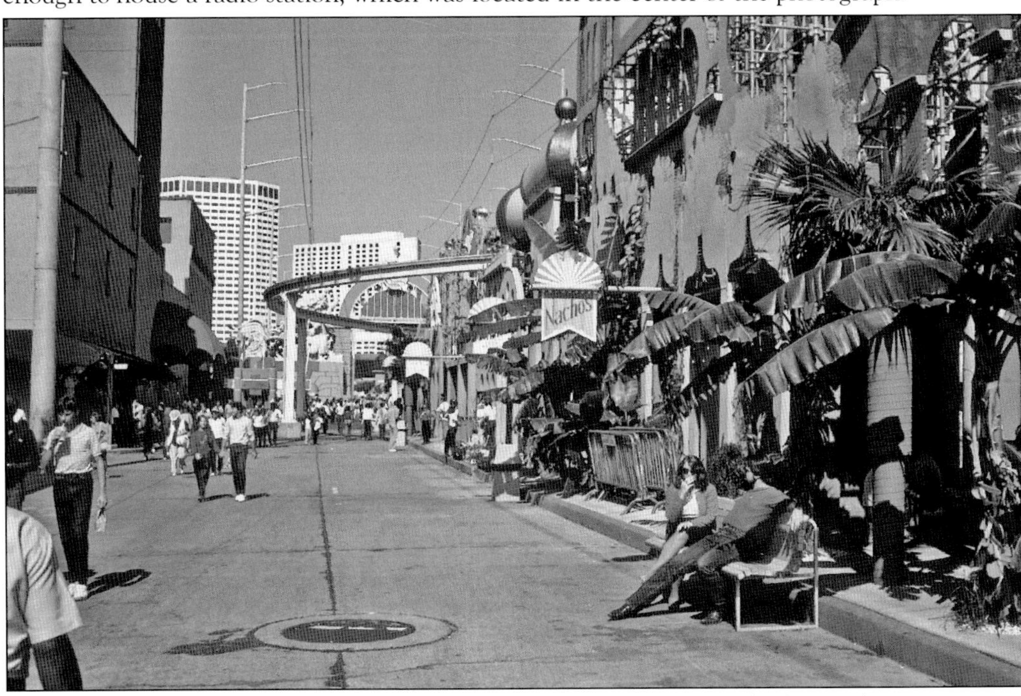

This section of the Wonderwall may look quite solid from this angle, but a closer inspection reveals the framework of the temporary structure. The power lines can clearly be seen on the left, but as intended, they are less obvious on the right due to the Wonderwall.

The Wonderwall was more than art and more than an intentional distraction. It housed a significant number of shops and snack bars, all of which hoped to tempt guests inside as they studied the unusual exterior.

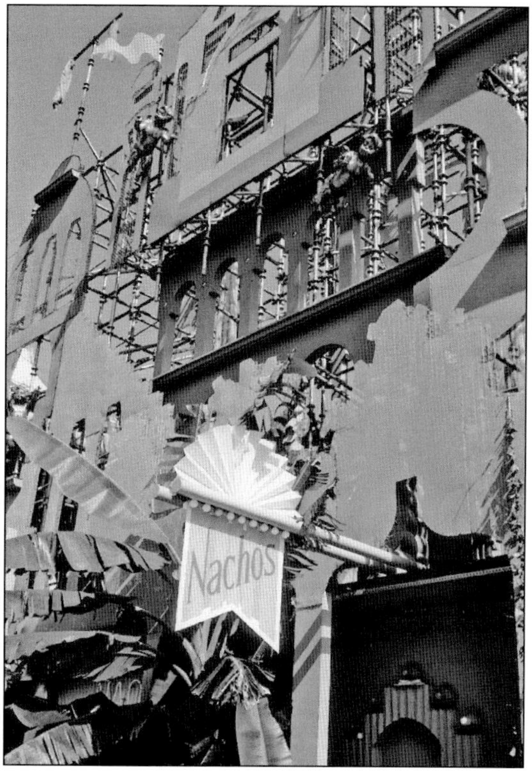

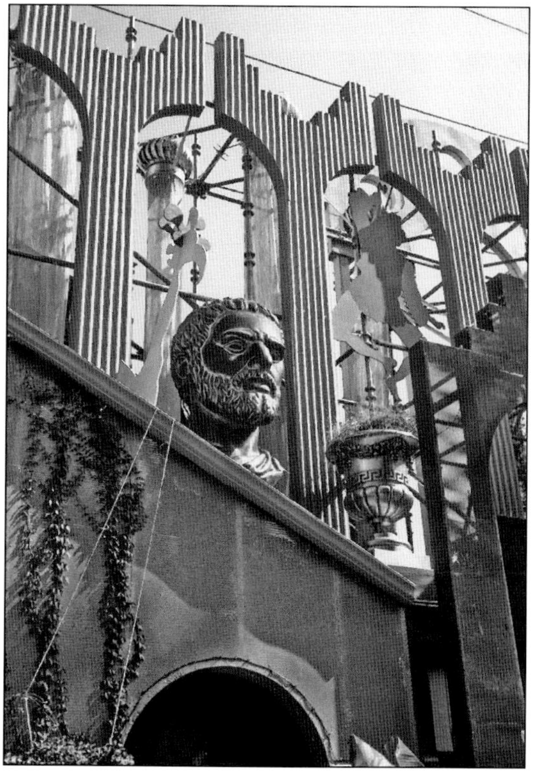

Many different art elements were set into the Wonderwall, including busts, urns, oversized peacocks, and columns, arches, and gazebos. Plants and fountains were interspersed to make the structure less formidable.

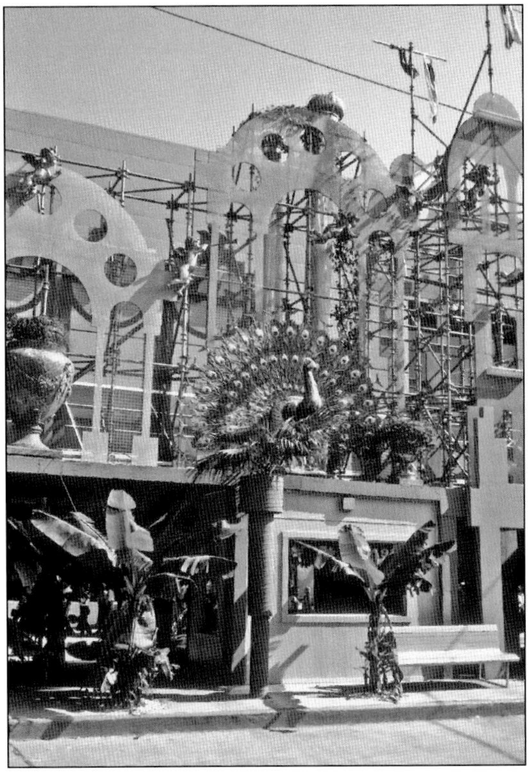

Among the shops and restaurants set into the Wonderwall was the studio of shortwave radio station WRNO. Operating daily from the fair site, disc jockeys offered to play requests from guests for their friends back home.

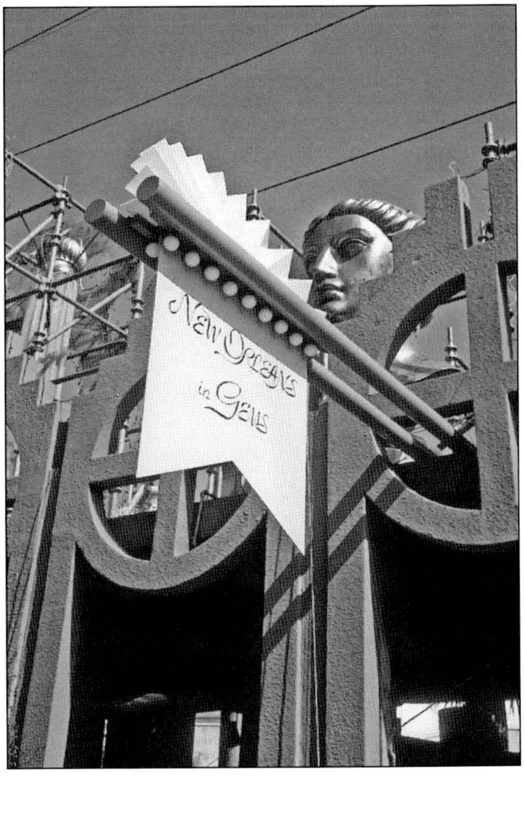

Visitors could purchase candy, cookies, ice cream, nachos, plants, toys, jewelry, logo merchandise, and gifts from India from shops inside the Wonderwall. For fun, children could even get their faces painted.

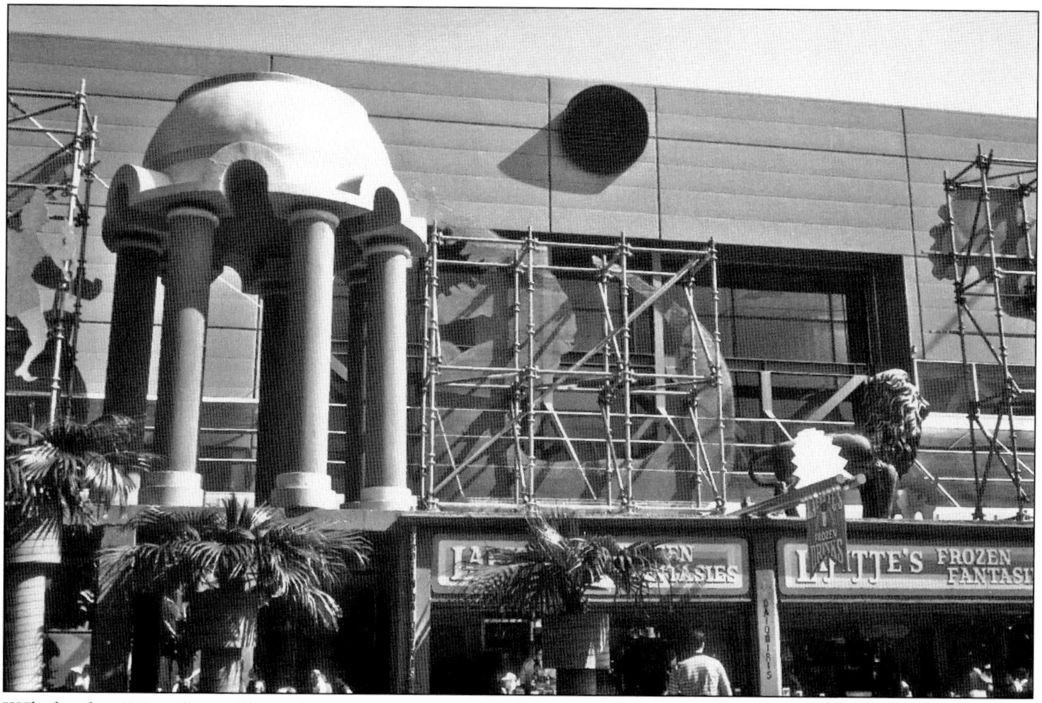

While the Wonderwall could not completely block out the structures on either side or the overhead power lines, it did succeed in drawing the eye away from them. The strange collection of shapes and colors was a visual delight, with no two sections alike.

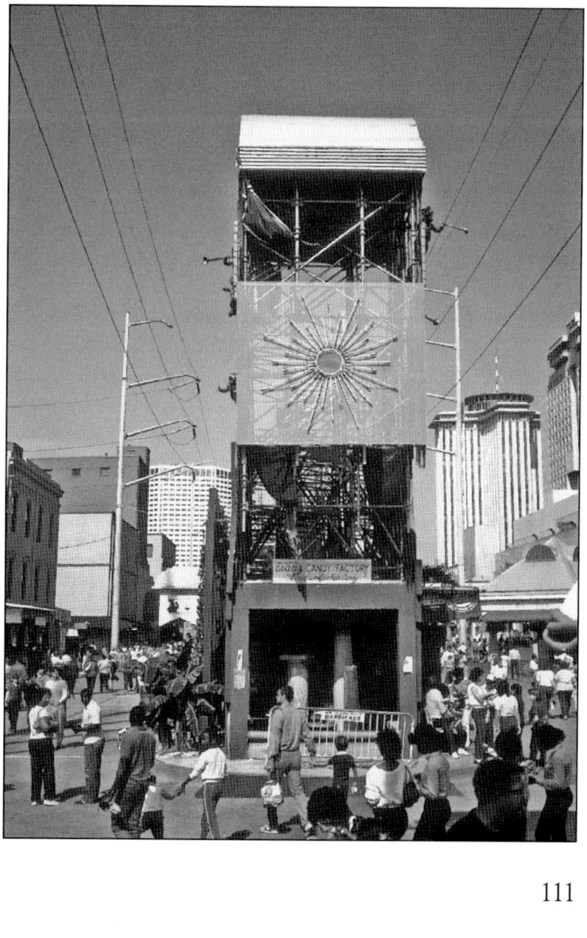

The size of the Wonderwall can be seen in this end view of one of the segments. While some parts of the fair failed to please architectural critics, the Wonderwall was well received and garnered a good deal of positive press.

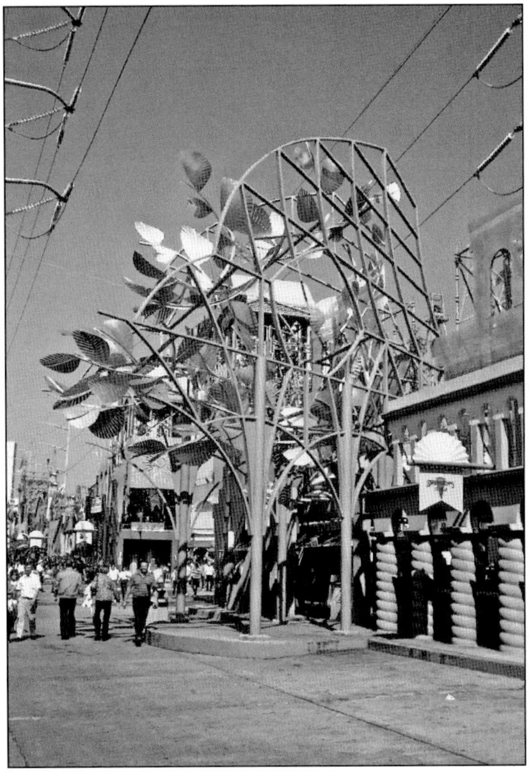

Tree Domes, a grove of metallic trees that stood 40 feet high with a span of 63 feet, was intended to inspire memories of shaded city streets, something that didn't exist at the fair.

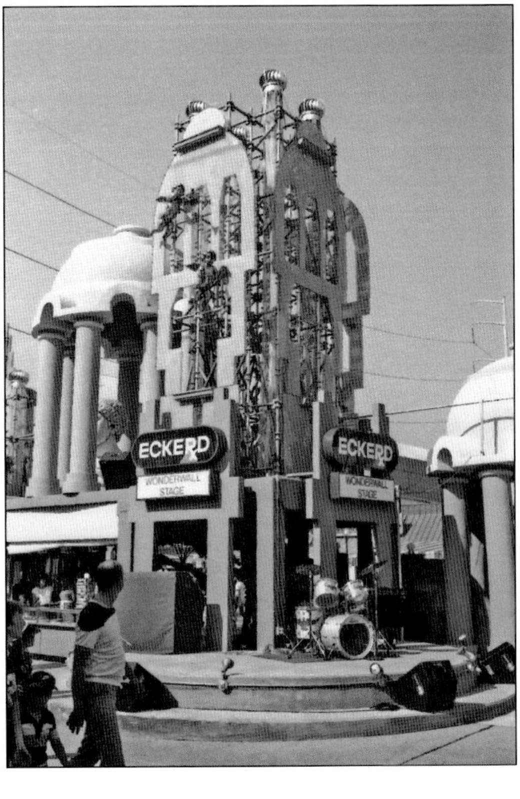

While at first glance this might be mistaken for a branch of a pharmacy chain of the time, the Wonderwall Stage was actually a small but lively entertainment area. Local performers were booked for a wide variety of acts that included jugglers, magicians, clowns, and musical groups.

Nine

Problems and Promises

It was the best of times; it was the worst of times

—Charles Dickens, *A Tale of Two Cities*

The 1984 New Orleans World's Fair was launched with grand hopes of bringing new business to the city and turning a profit for its investors. Designed under the assumption that it would attract 70,000 visitors per day, the fair looked like it might well succeed when 83,000 people attended the opening day. After that, though, the attendance hovered around 35,000 for the rest of the season, leaving the fair and its investors in the red.

With estimated losses over $100 million, it was inevitable that fingers would be pointed and lawsuits filed. Many theories were put forth as to what went wrong, with charges that ranged from incompetence to intentional fraud. Studies conducted as part of the lawsuits eventually found that the attendance estimates had been based on flawed tourism data that over-counted the state's tourism business. Because of that mistake, the fair was effectively doomed from the start.

Other errors contributed to the fiscal mess. Perhaps the largest was the decision to push the fair back from its original 1982 date, for that allowed Knoxville to steal the spotlight with its fair, and potential ticket buyers then had to choose between two relatively close events. When the Knoxville fair was panned, they may have just decided to skip both fairs. Possibly even worse was the 1982 opening of Walt Disney World's Epcot Center, which was touted as a permanent world's fair. Some studies suggested that the 1984 Olympics in Los Angeles further muddied the situation.

Whatever the reasons, the fair lost money. Branded as a failure, the fair and its organizers were soundly criticized for years before it faded into obscurity. Then, about 20 years later, writers describing the attractions of the city's Arts District began writing about how the fair was the forgotten catalyst for the area's rebirth. Finally it was seen that the fair had done what it was supposed to do. It helped launch the convention center, a cruise ship terminal, and a shopping mall. It was the impetus for new hotels, shops, and restaurants built around the increased convention business. It also left many people with happy memories of the year the world came to New Orleans.

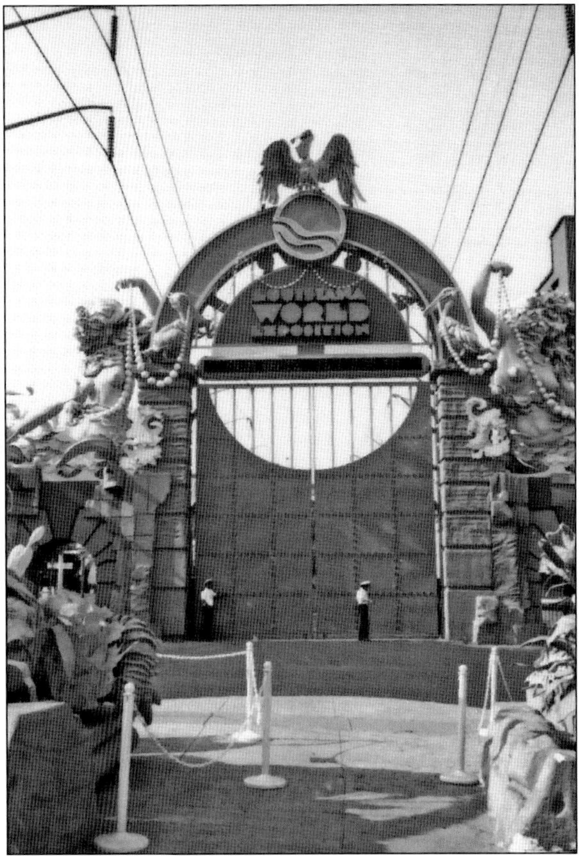

All good things come to an end, and on November 11, 1984, the gates swung shut for the final time. More than 7 million people had visited the fair, but the project had been financed and built with a goal of 12 million visitors. Seriously short on operating funds as a result, the fair had to declare bankruptcy while still operating, the only world's fair to do so.

The biggest problem the fair had was a lack of visitors. While those who attended seemed to enjoy it, a combination of factors led to far fewer guests than had been projected. That in turn led to a severe shortfall in the funds needed to operate the fair and to repay investors. Scenes like this of ticket booths without lines were far too common.

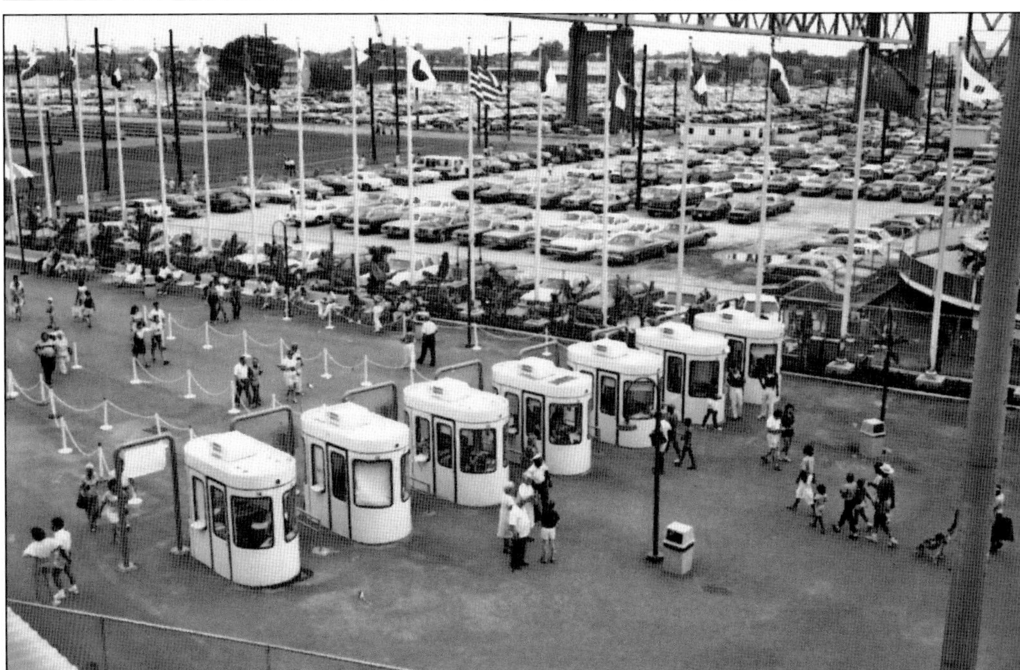

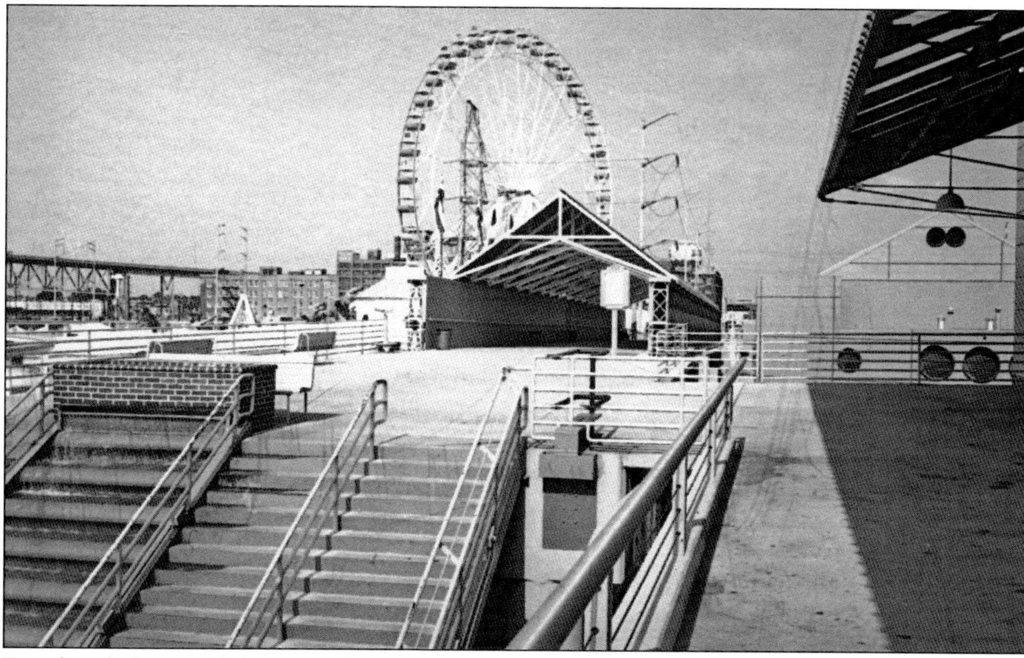

Faced with fewer exhibitors and sponsors than originally projected, the fair had been forced to cancel some projects and scale back others. The Crystal Palace, to be built entirely of glass, was one pavilion that never made it past the early stages of the design process. This open area at the end of the International Riverfront was originally to have been called Celebration Plaza, but when finished, it was little more than an empty expanse of concrete.

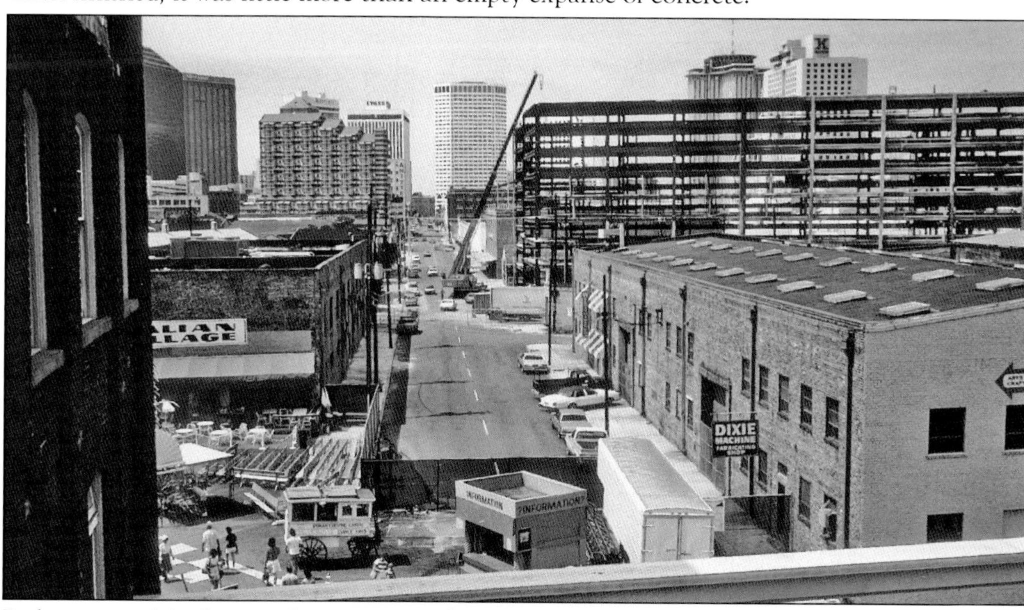

Perhaps one of the biggest shortcomings of the fair was the ever-present intrusion of the real world just outside the fences. While the fair designers succeeded in creating a unique setting across most of the grounds, with fanciful buildings and bright colors to entertain the eye, the old warehouses and construction sites in the rest of the city presented a jarring counterpart that unfortunately detracted from the fair experience.

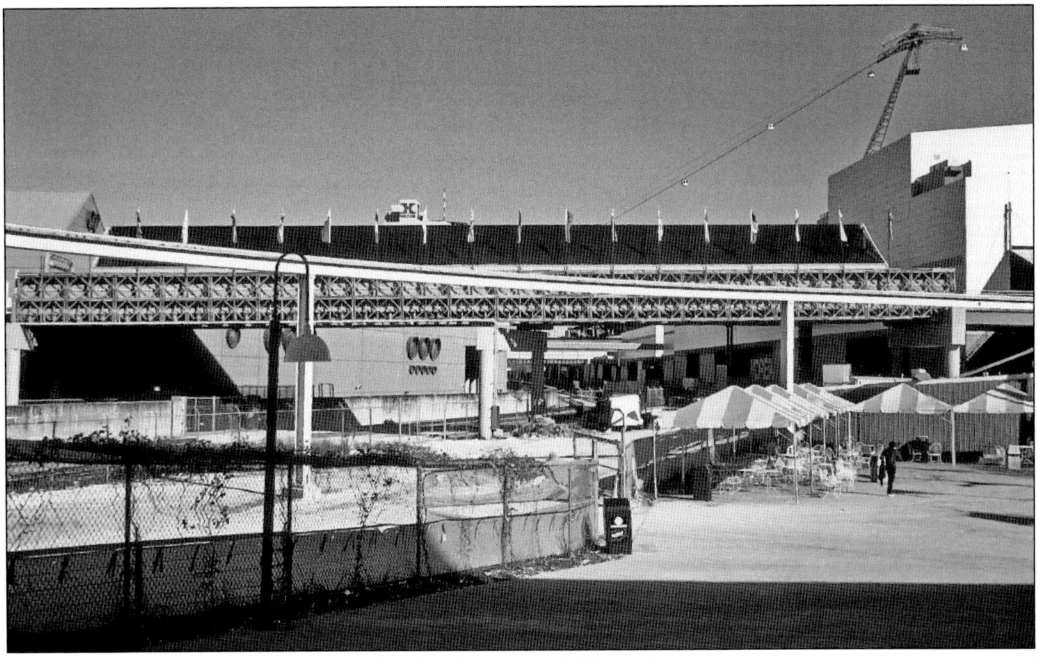

Not all of the problems were outside the fair gates. When the money ran out, some sections of the site could not be completed as planned. This view from the United States Pavilion shows how even basic things like the perimeter fencing were left in a rather unappealing state, with not even a hint of landscaping anywhere in sight.

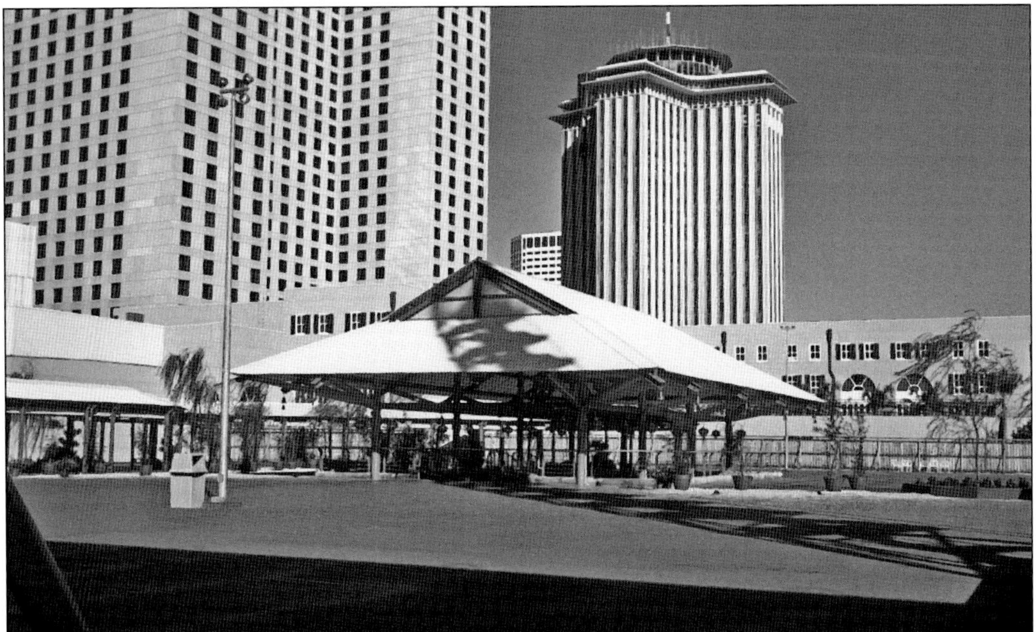

The addition of landscaping was not necessarily much of an improvement. This area at the top of the China Pavilion included a number of potted trees, but there was nothing else there to entertain visitors. Despite its financial problems, the grounds were kept clean, partially thanks to prisoners from local jails who worked at night during the last weeks of the fair.

Many of the vendors who once had high hopes for huge profits found the fair to be a disappointing—and costly—experience. For example, with millions of visitors predicted, it must have seemed like a safe bet to print 200,000 copies of the official guidebook, with more editions sure to follow. In reality, only 25,000 copies were ever sold, with the remainder sadly sent to a landfill.

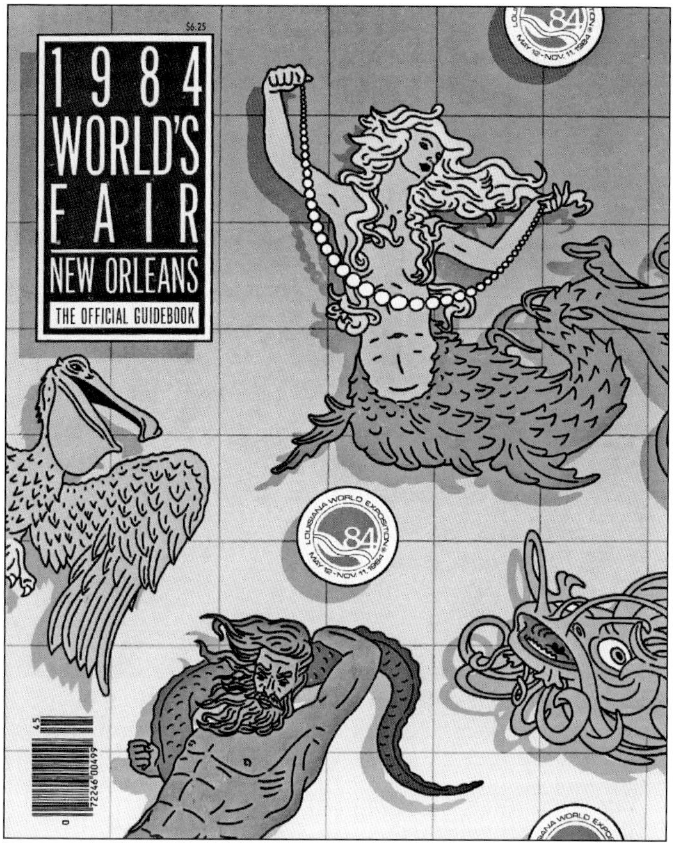

A few people did make a profit off the fair, for there were some real bargains for those bidding on the fair assets after the gates closed. The $5.5-million United States Pavilion sold for just $130,000 to a developer who hoped to find a new home for the steel-framed building. It is unknown if the building was ever rebuilt in a new location.

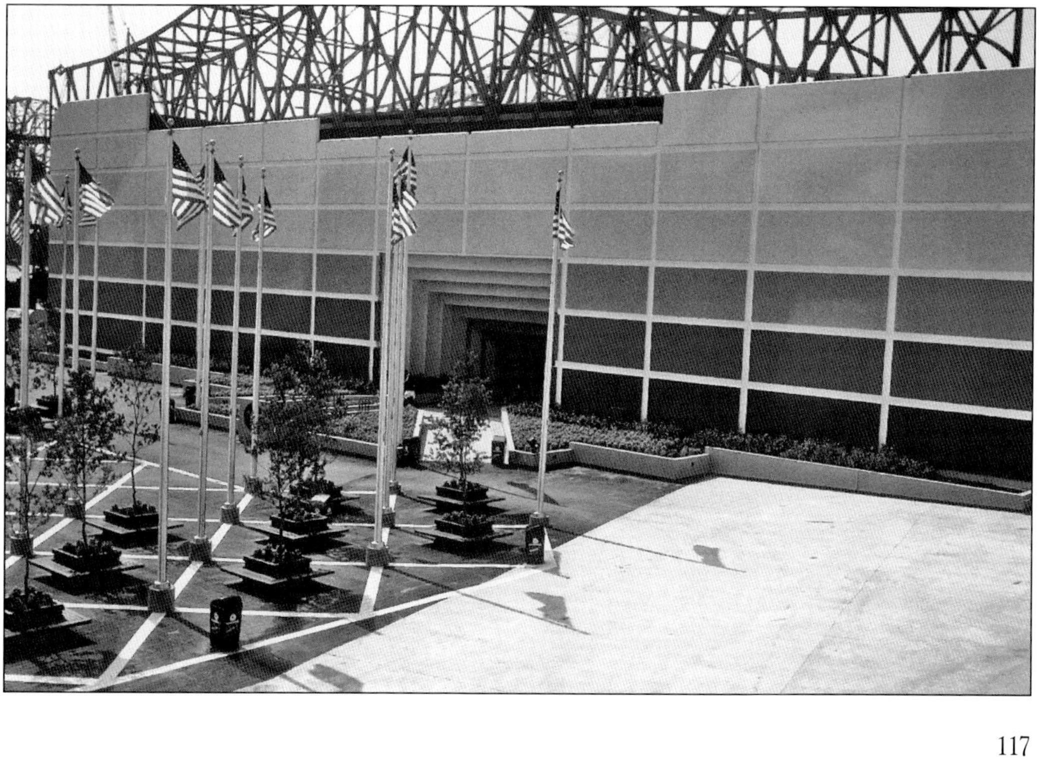

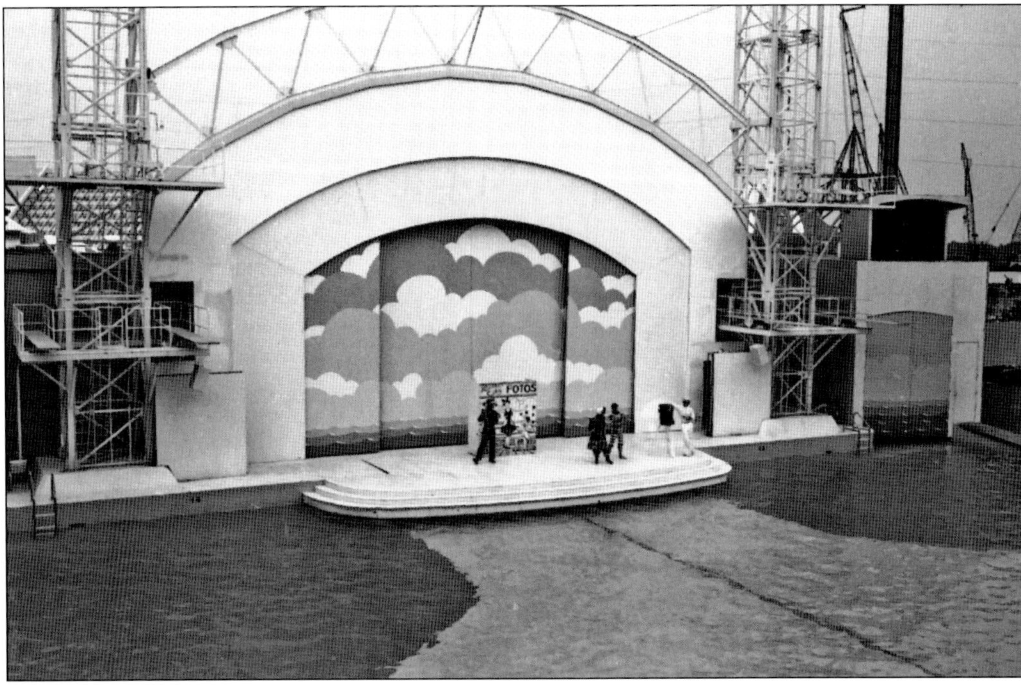

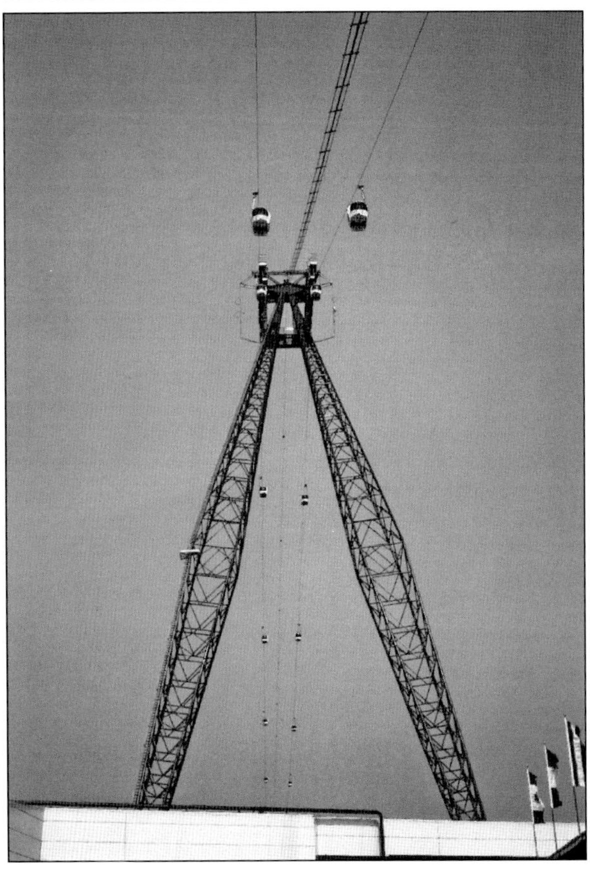

Some fair structures were bought for just for their scrap value, but others were acquired for purposes quite different than what their designers had originally intended. The main shell of the Aquacade, for example, was purchased for reuse at a local cemetery. Like the United States Pavilion, it is unknown if the structure was reassembled.

The MART system stayed open when the fair ended, for it had been planned as a permanent transportation system for local commuters. The hoped-for ridership didn't materialize, though, and the system was shut down in April 1985. Despite several attempts to reuse or relocate the system, it continued to languish until it was removed in 1994 after the U.S. Coast Guard decided it was a potential hazard to navigation.

The critically acclaimed Wonderwall was broken apart when the fair was demolished. While some of the artwork was returned to the original designers, many pieces of it were sold at auction. Some of these remnants continue to surface at art dealers or auctions today.

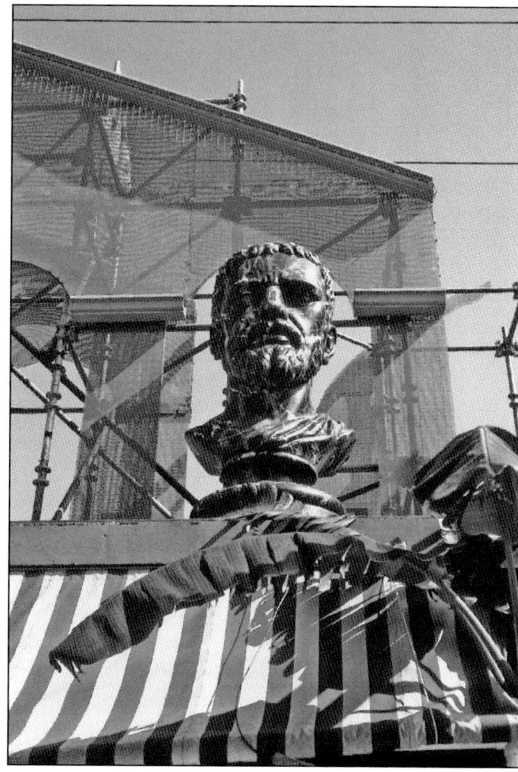

Some of the artwork assembled for the fair is still on public display, such as *The Source*, an entry in the fair's International Water Sculpture Competition. It can be seen today near Elk Place and Canal Street but in a rather less attractive setting and without the original fountain components. (Courtesy of Mike Hoffman.)

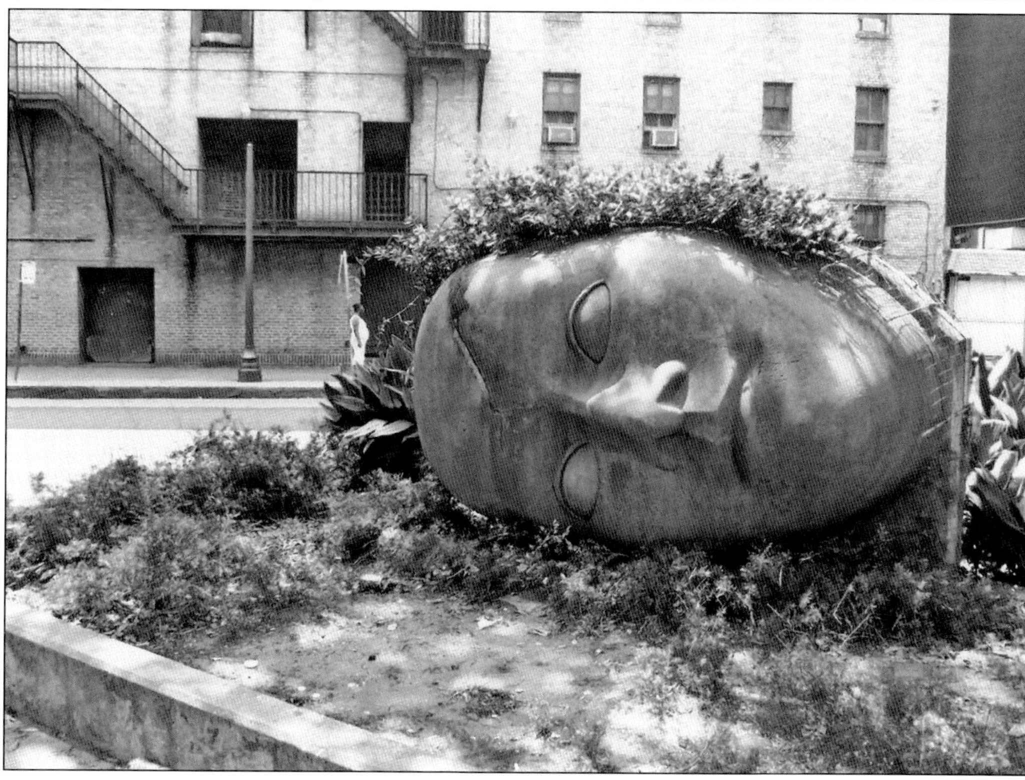

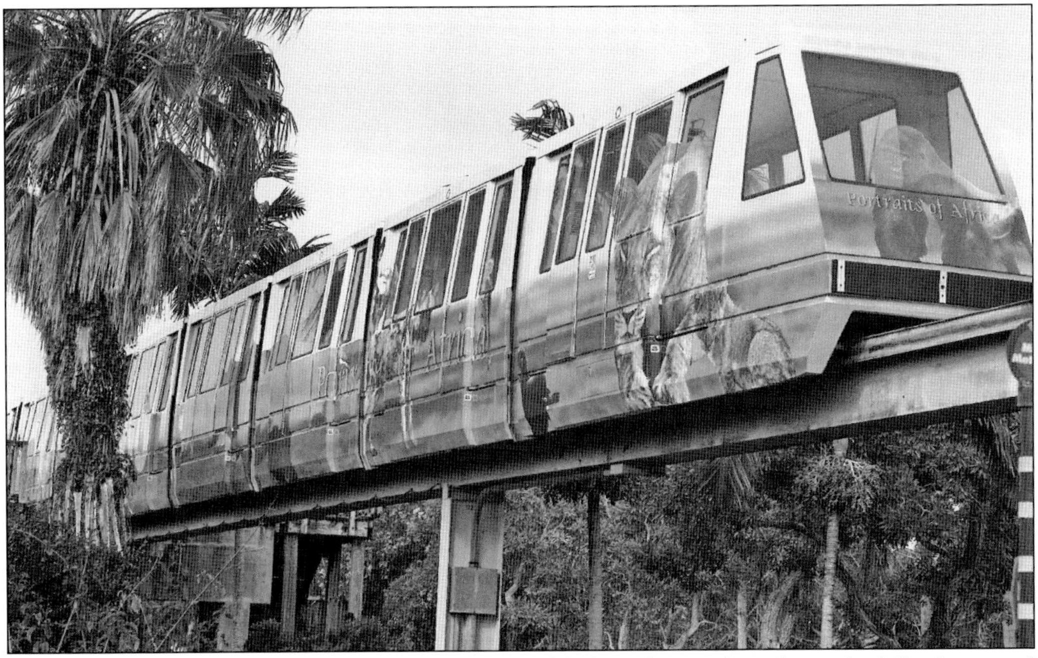

Most visitors to the Miami Dade Metrozoo are unlikely to know that this colorfully decorated monorail once toured the fair site. Moved to the zoo at the closing of the fair, the system was damaged by Hurricane Andrew, but it has been restored and is still in daily use, making it one of the longest-running monorail systems in the United States. (Courtesy of Ron Magill.)

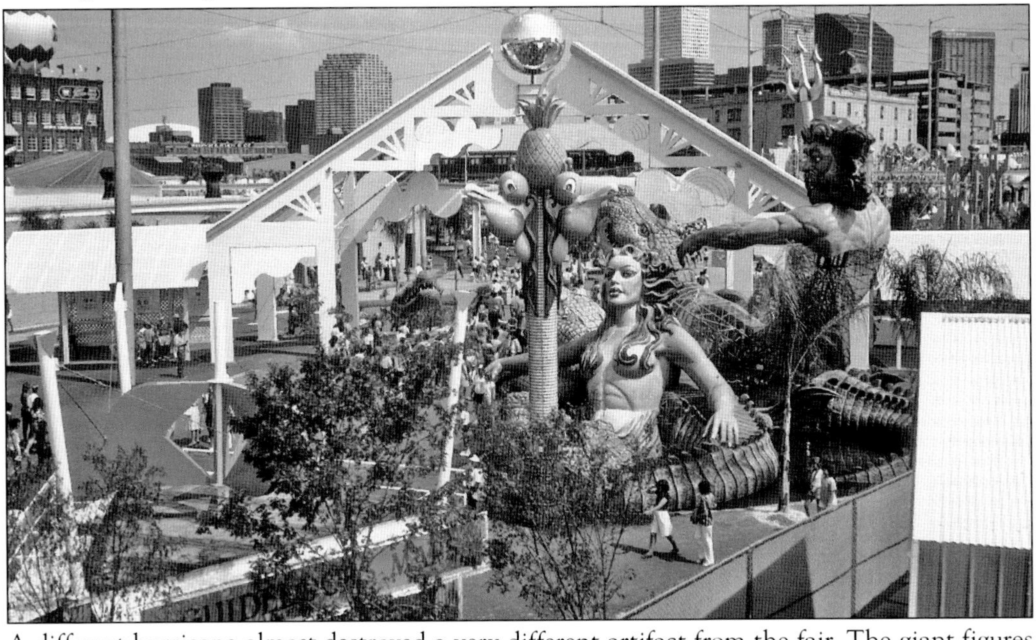

A different hurricane almost destroyed a very different artifact from the fair. The giant figures that had adorned the Bridge Gate entrance had been disassembled and stored for decades in a warehouse, later to be threatened by the flooding of Hurricane Katrina. Luckily they survived, and plans are underway to display them in an expansion of designer Blaine Kern's Mardi Gras World attraction.

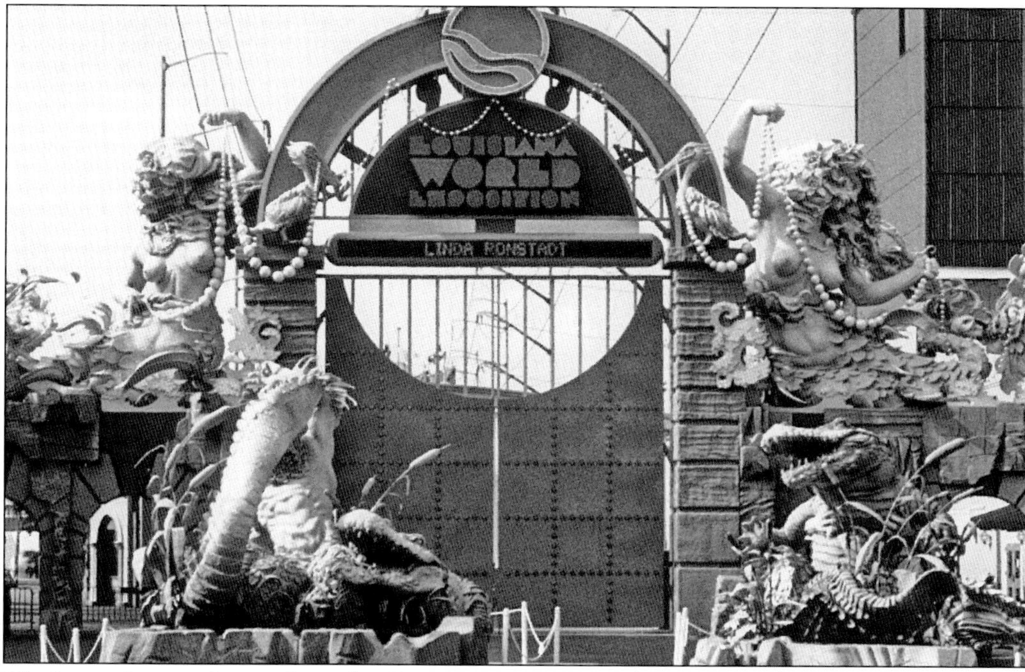

The better-known figures from the City Gate were not as fortunate, and the controversial bare-chested mermaids are no longer with us, having been taken apart after the fair. In a bit of whimsy, though, designer Joe Barth did incorporate one of the giant pearls into a design for a restaurant in nearby Slidell.

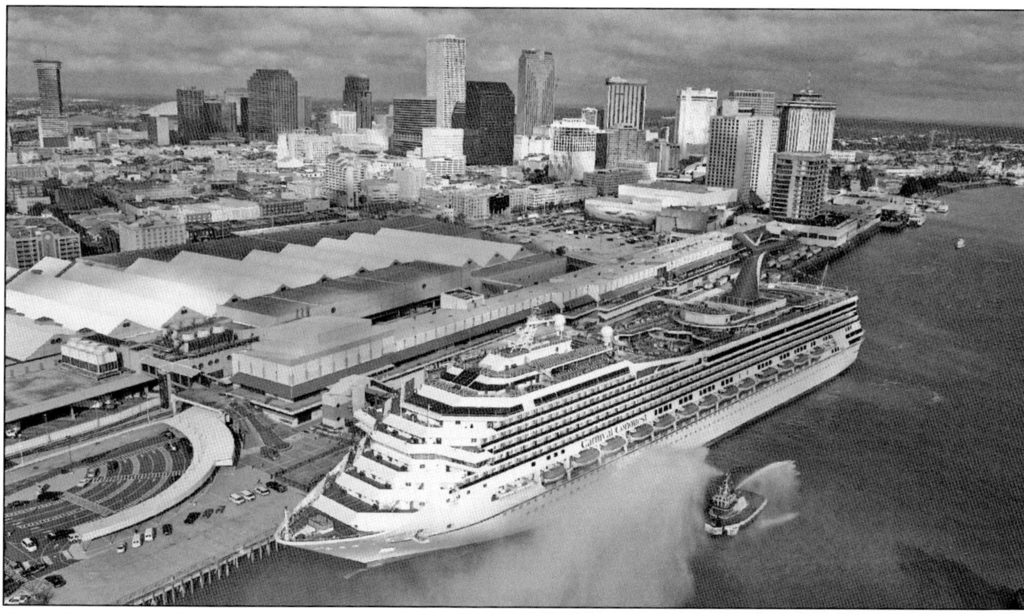
A major part of the fair can be seen today in a very different role. Most of the former International Riverfront building now serves as the Julia Street Cruise Terminal. The rest of the Riverfront area was turned into the Riverwalk, a mall and hotel complex. Although both were damaged by Hurricane Katrina, they have been repaired and are two important legacies of the fair. (Courtesy of Donn Young/Port of New Orleans.)

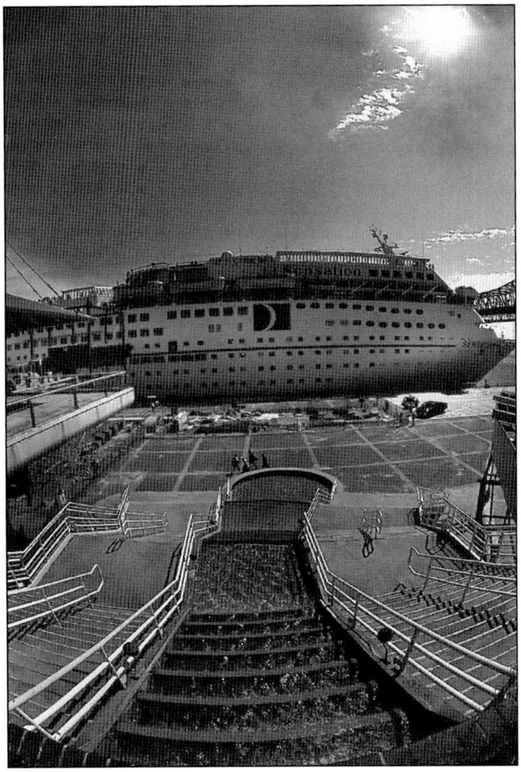

The staircase and fountain at the end of the International Riverfront building are still there. While the steps don't really lead anywhere now, the fountain has been restored and still cools the feet of visitors to the area. (Courtesy of Donn Young/Port of New Orleans.)

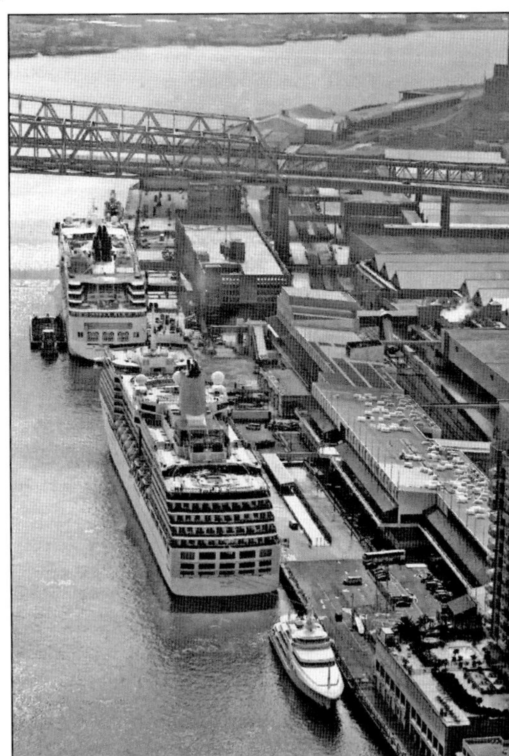

Just as the fair planners had hoped, the formerly derelict port area has become a major source of tourist revenue for the city. A new terminal was recently built on the site of the United States Pavilion next to the Julia Street terminal. Smaller boats can dock at the Riverwalk where the China Pavilion once stood. (Courtesy of Donn Young/Port of New Orleans.)

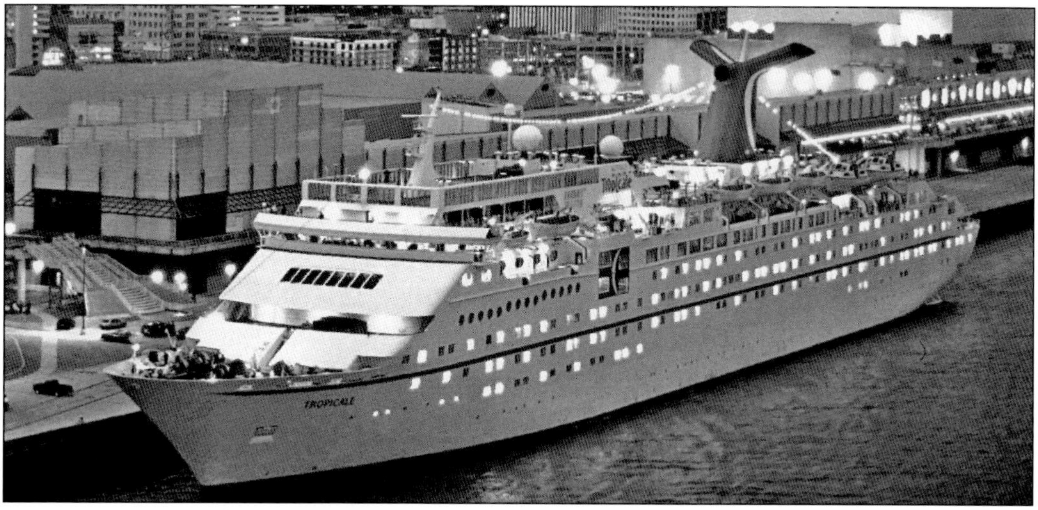

Sometimes parts of the fair show up in unexpected ways. The building behind the cruise ship was the Canada Pavilion, and if one looks closely near the top, one can still see part of the Canadian flag that once adorned the building. The former pavilion was to have become a jazz complex, but the project collapsed, and the building has never been fully utilized since the fair. (Courtesy of Donn Young/Port of New Orleans.)

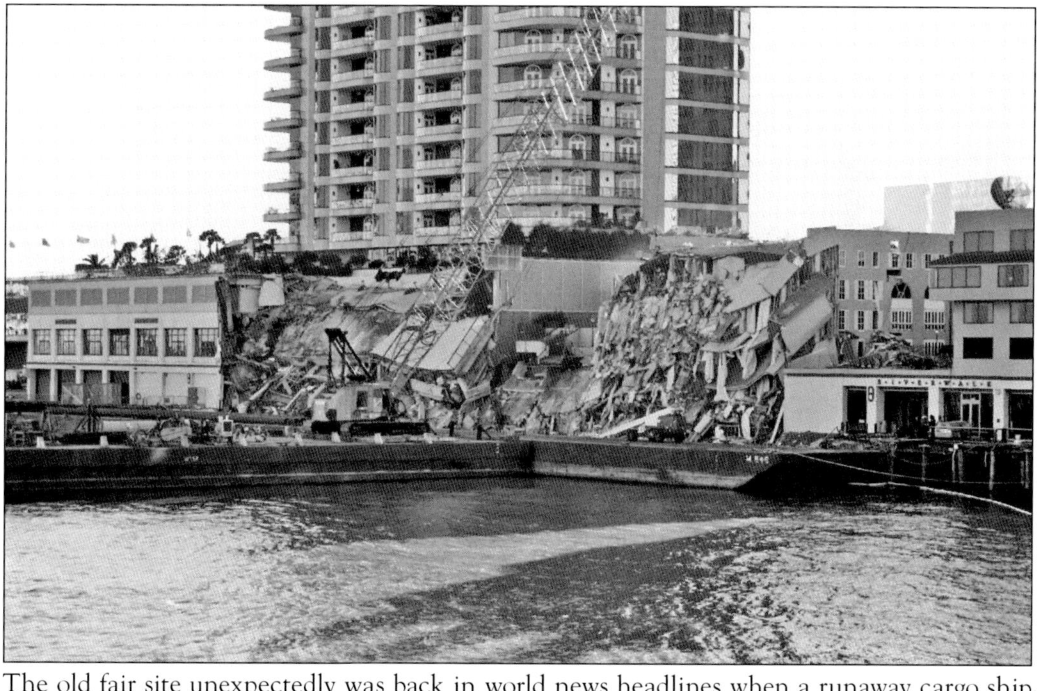

The old fair site unexpectedly was back in world news headlines when a runaway cargo ship collided with the Riverwalk mall on December 14, 1996. The MV *Bright Star* lost steering control and hit the mall, costing an estimated $15 million in damages. While 66 people were injured, no one was killed in the incident. (Courtesy of Lee Taylor.)

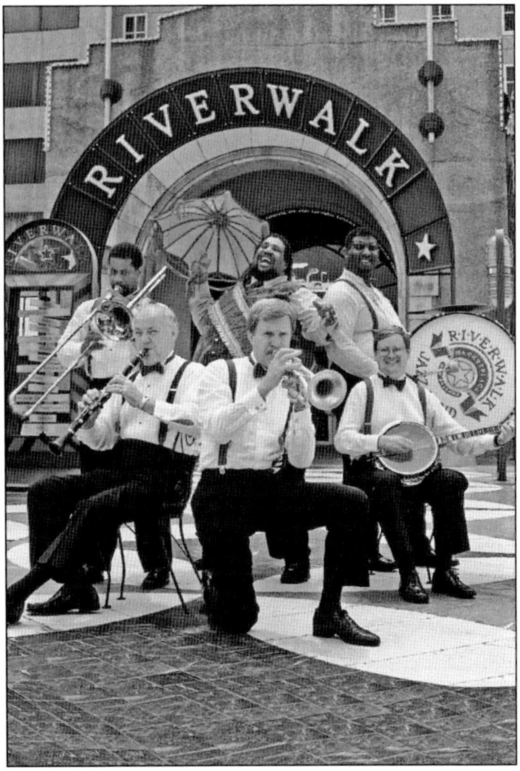

The northern end of the International Riverfront was developed into the Riverside Marketplace, a three-story shopping mall and hotel complex. The mall, which was part of the fair's master plan, opened on August 28, 1986. (Courtesy of Jackson Hill Photography.)

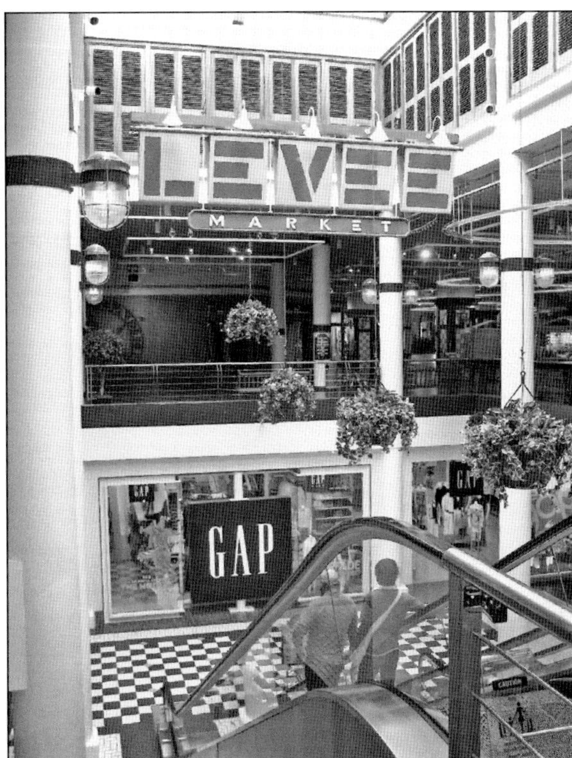

Like much of New Orleans, the Riverwalk facility was damaged during Hurricane Katrina and the looting that followed. The mall has since been fully repaired and now features 140 shops and restaurants. (Courtesy of Jackson Hill Photography.)

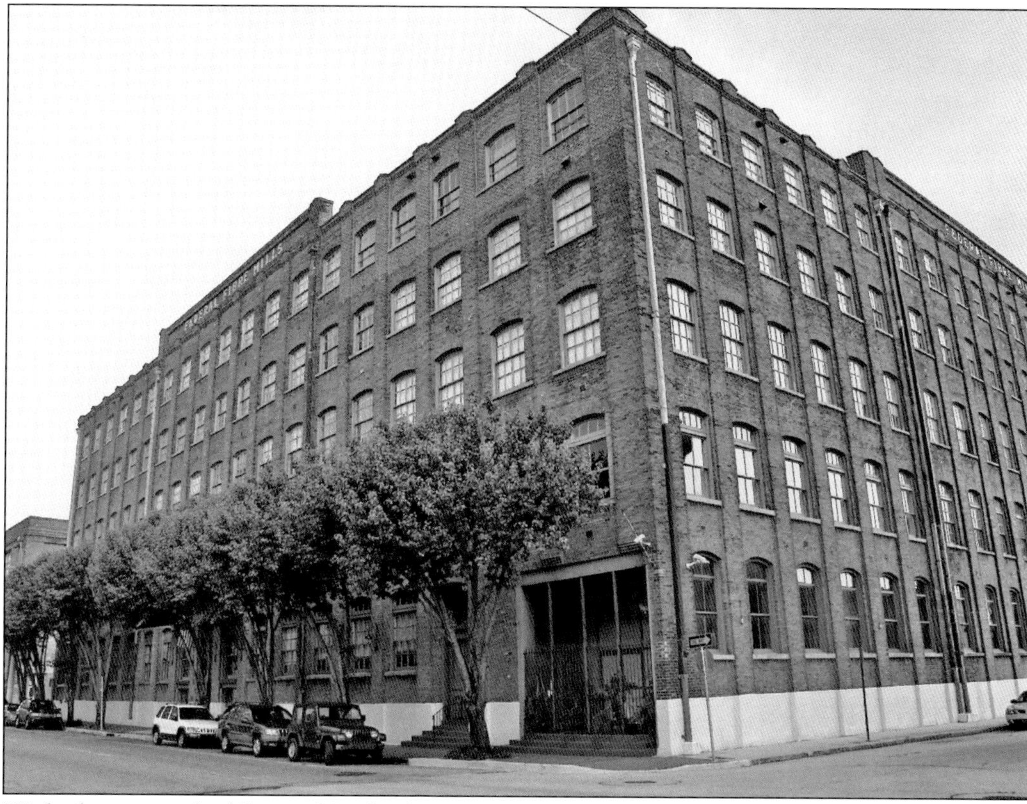

With the removal of Festival Park, the majority of the area is now a welcome oasis of green. The Federal Fibre Mills Building, which once presided over Festival Park, now looks down on a city park. The building itself has been converted into prestigious condominiums. (Courtesy of Eric Bouler.)

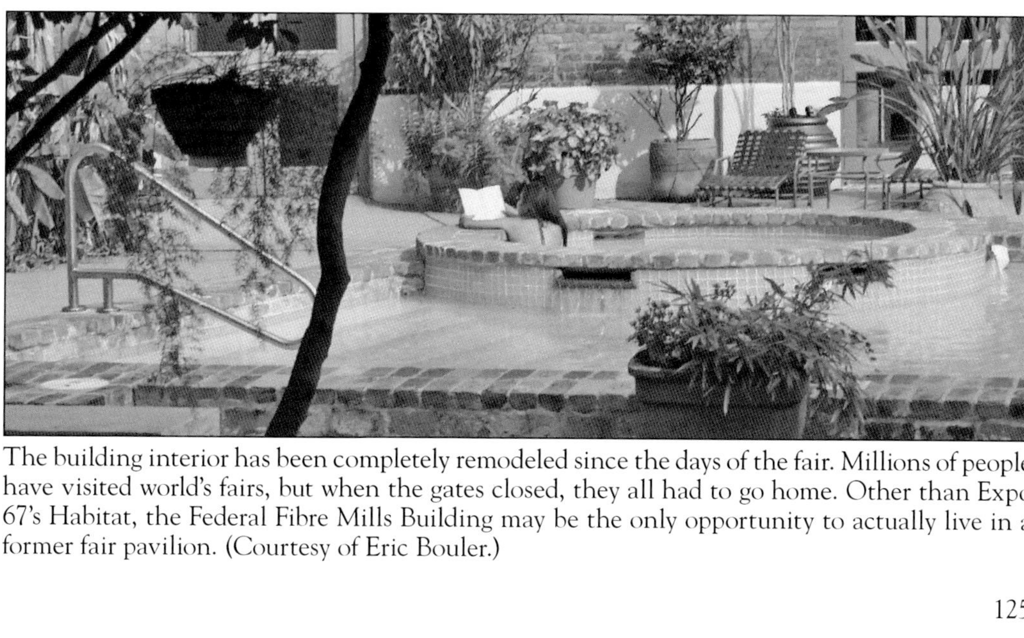

The building interior has been completely remodeled since the days of the fair. Millions of people have visited world's fairs, but when the gates closed, they all had to go home. Other than Expo 67's Habitat, the Federal Fibre Mills Building may be the only opportunity to actually live in a former fair pavilion. (Courtesy of Eric Bouler.)

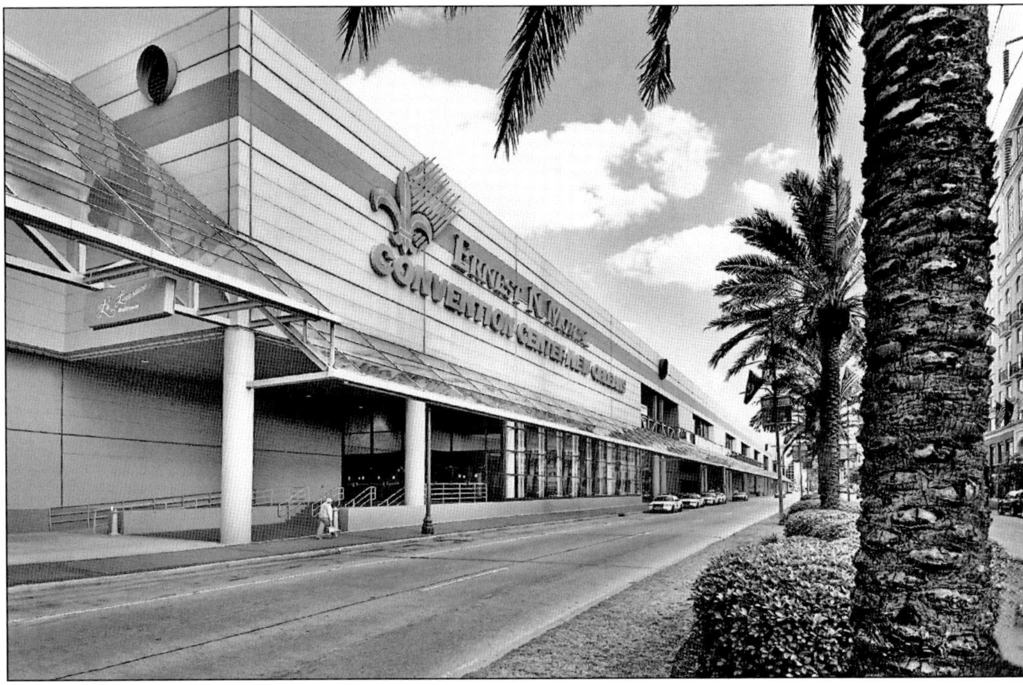

The Wonderwall is long gone, replaced now by palm trees, and the signs for the Louisiana Journey are but memories. The Julia Street entrance to the convention center still looks very much like it did during the building's tenure as the Great Hall though, making it possible to visualize where the rest of the fair once stood. (Courtesy of the Ernest N. Morial Convention Center.)

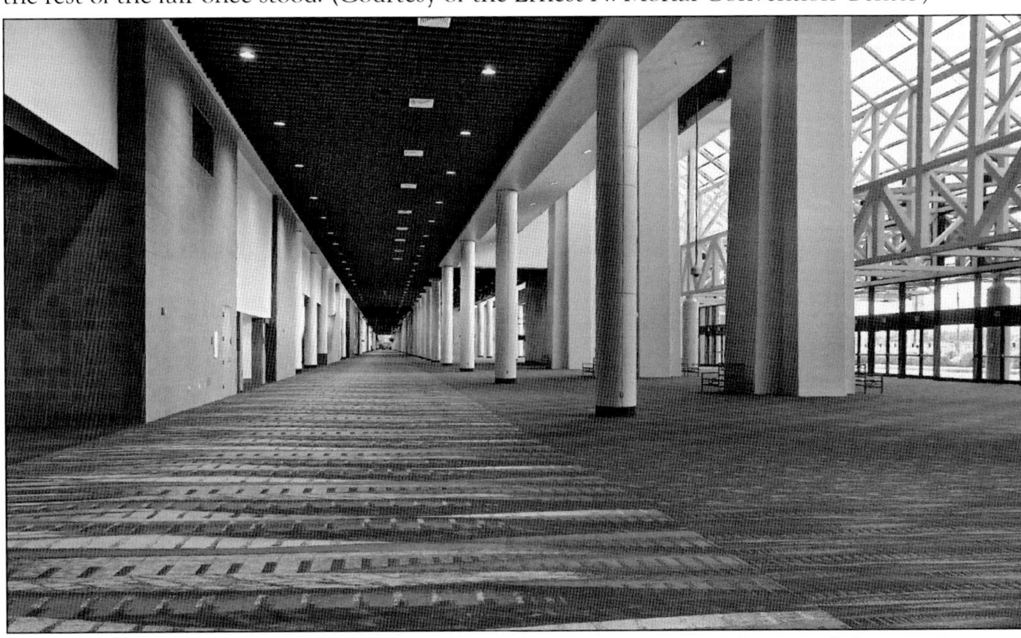

In the days right after Hurricane Katrina, this area was a scene of chaos and despair, for it was one of the major centers used to house the thousands of people displaced by the flooding that devastated so much of the city. Although never planned for such a use, the convention center was crucial in the initial recovery efforts. (Courtesy of the Ernest N. Morial Convention Center.)

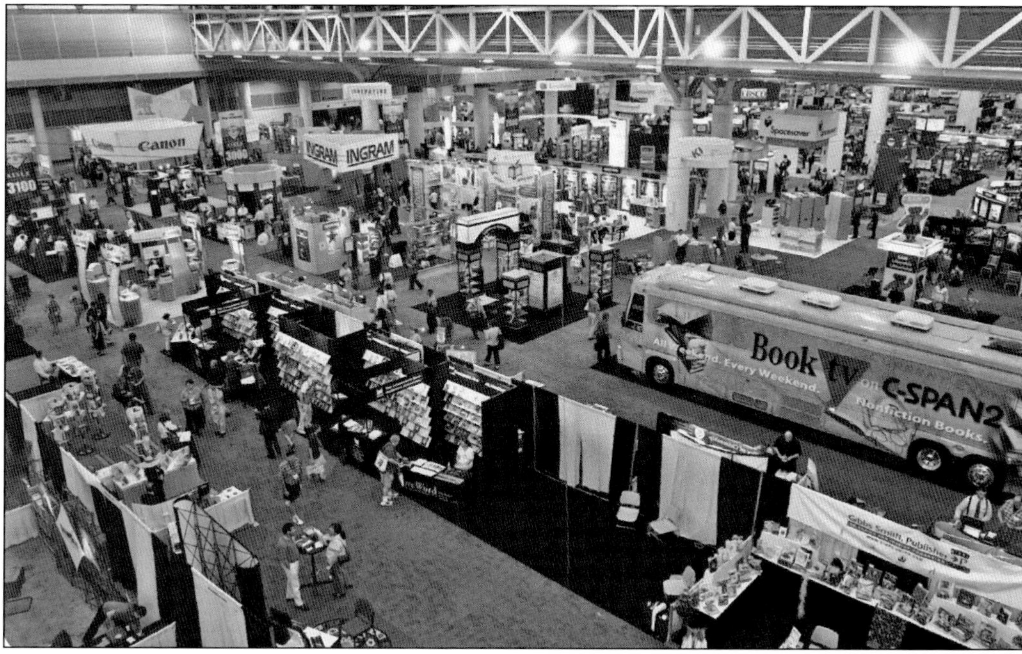

When the refugees left, the center needed a thorough cleaning and repair, a task that took 13 months. This post-Katrina view shows how the size of the former Great Hall is well suited to the large conventions the original planners had hoped to attract. (Courtesy of the Ernest N. Morial Convention Center.)

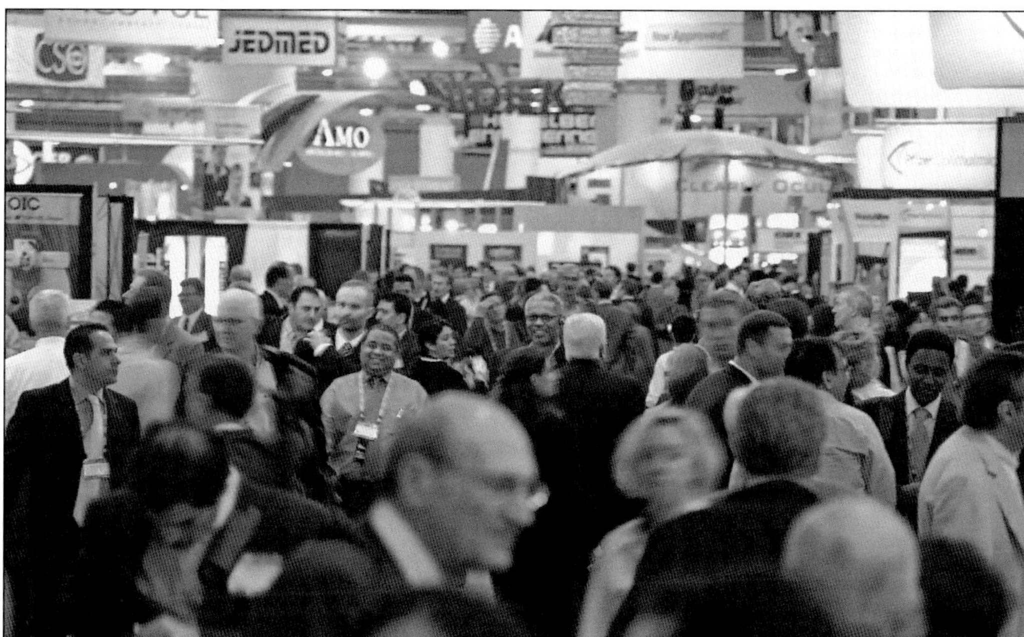

In its first 20 years of operation, the center hosted more than 1,600 events, bringing an estimated 10.6 million visitors to the city. Perhaps most importantly, it generated an estimated $37 billion in revenue—bringing tourist dollars into the state just as the fair planners had hoped in 1976, back when it all started. (Courtesy of the Ernest N. Morial Convention Center.)

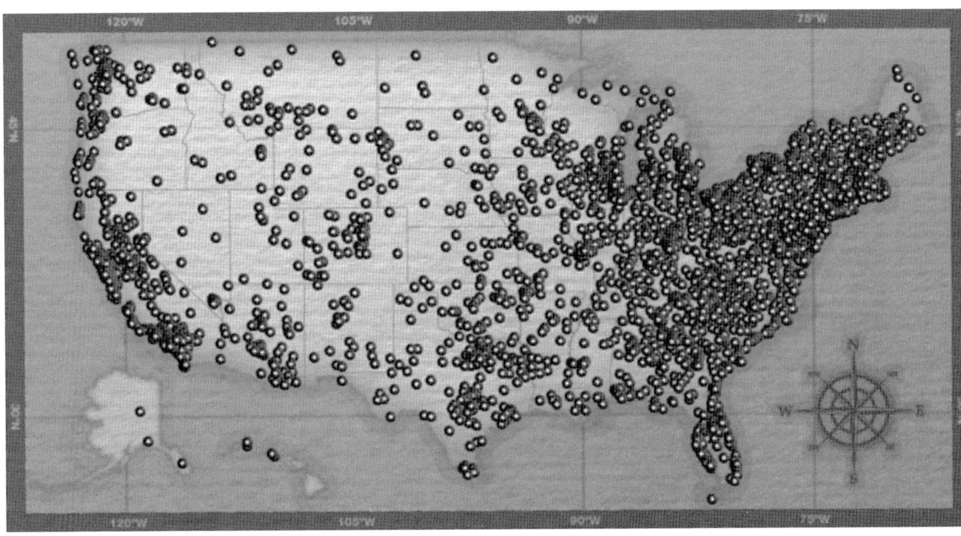